P9-BJJ-091

SOURCE
The Prentice Hall
ENGINEERING SOURCE

# Design Concepts for Engineers

## Second Edition

## Mark N. Horenstein

*Boston University*

Prentice Hall
Upper Saddle River, NJ 07458

**Library of Congress Cataloging-in-Publication Data**

Horenstein, Mark
    Design Concepts for Engineers / Mark N. Horenstein,—2nd ed.
      p. cm. — (ESource—the Prentice Hall engineering source)
    Includes bibliographical references and index.
    ISBN 0–13–093430-5
    1.  Engineering—Study and teaching.  2.  Engineering students.  I.  Title.  II.  Series.

T65 .S3858 2001
620'.0071'1—dc21

2001028829

Vice President and Editorial Director, ECS: **Marcia J. Horton**
Executive Editor: **Eric Svendsen**
Associate Editor: **Dee Bernhard**
Vice President and Director of Production and Manufacturing, ESM: **David W. Riccardi**
Executive Managing Editor: **Vince O'Brien**
Managing Editor: **David A. George**
Production Editor: **Scott Disanno**
Director of Creative Services: **Paul Belfanti**
Creative Director: **Carole Anson**
Art Director: **Jayne Conte**
Art Editor: **Adam Velthaus**
Manufacturing Manager: **Trudy Pisciotti**
Manufacturing Buyer: **Lisa McDowell**
Marketing Manager: **Holly Stark**
Marketing Assistant: **Karen Moon**

© 2002 by Prentice-Hall, Inc.
Upper Saddle River, New Jersey 07458

10  9  8  7  6  5  4  3  2

ISBN 0-13-093430-5

Prentice-Hall International (UK) Limited, *London*
Prentice-Hall of Australia Pty. Limited, *Sydney*
Prentice-Hall Canada, Inc., Toronto
Prentice-Hall Hispanoamericana, S.A., *Mexico City*
Prentice-Hall of India Private Limited, *New Delhi*
Prentice-Hall of Japan, Inc., *Tokyo*
Pearson Education (Singapore) Pte. Ltd., *Singapore*
Editoria Prentice-Hall do Brasil, Ltda., *Rio de Janeiro*

# About ESource

## ESource—The Prentice Hall Engineering Source
## —www.prenhall.com/esource

ESource—The Prentice Hall Engineering Source gives professors the power to harness the full potential of their text and their first-year engineering course. More than just a collection of books, ESource is a unique publishing system revolving around the ESource website—www.prenhall.com/esource. ESource enables you to put your stamp on your book just as you do your course. It lets you:

*Control* You choose exactly what chapter or sections are in your book and in what order they appear. Of course, you can choose the entire book if you'd like and stay with the authors' original order.

*Optimize* Get the most from your book and your course. ESource lets you produce the optimal text for your students needs.

*Customize* You can add your own material anywhere in your text's presentation, and your final product will arrive at your bookstore as a professionally formatted text. Of course, all titles in this series are available as stand-alone texts, or as bundles of two or more books sold at a discount. Contact your PH sales rep for discount information.

## ESource ACCESS

Professors who choose to bundle two or more texts from the ESource series for their class, or use an ESource custom book will be providing their students with complete access to the library of ESource content. All bundles and custom books will come with a student password that gives web ESource ACCESS to all information on the site. This passcode is free and is valid for one year after initial log-on. We've designed ESource ACCESS to provide students a flexible, searchable, on-line resource. Professors may also choose to deliver custom ESource content via the web only using ESource ACCESS passcodes. Contact your PH sales rep for more information.

## ESource Content

All the content in ESource was written by educators specifically for freshman/first-year students. Authors tried to strike a balanced level of presentation, an approach that was neither formulaic nor trivial, and one that did not focus too heavily on advanced topics that most introductory students do not encounter until later classes. Because many professors do not have extensive time to cover these topics in the classroom, authors prepared each text with the idea that many students would use it for self-instruction and independent study. Students should be able to use this content to learn the software tool or subject on their own.

While authors had the freedom to write texts in a style appropriate to their particular subject, all followed certain guidelines created to promote a consistency that makes students comfortable. Namely, every chapter opens with a clear set of **Objectives**, includes **Practice Boxes** throughout the chapter, and ends with a number of **Problems**, and a list of **Key Terms. Applications Boxes** are spread throughout the book with the intent of giving students a real-world perspective of engineering. **Success Boxes** provide the student with advice about college study skills, and help students avoid the common pitfalls of first-year students. In addition, this series contains an entire book titled **Engineering Success** by Peter Schiavone of the University of Alberta intended to expose students quickly to what it takes to be an engineering student.

## Creating Your Book

Using ESource is simple. You preview the content either on-line or through examination copies of the books you can request on-line, from your PH sales rep, or by calling 1-800-526-0485. Create an on-line outline of the content you want, in the order you want, using ESource's simple interface. Either type or cut and paste your own material and insert it into the text flow. You can preview the overall organization of the text you've created at anytime (please note, since this preview is immediate, it comes unformatted.), then press another button and receive an order number for your own custom book. If you are not ready to order, do nothing—ESource will save your work. You can come back at any time and change, re-arrange, or add more material to your creation. Once you're finished and you have an ISBN, give it to your bookstore and your book will arrive on their shelves four to six weeks after they order. Your custom desk copies with their instructor supplements will arrive at your address at the same time.

To learn more about this new system for creating the perfect textbook, go to www.prenhall.com/esource. You can either go through the on-line walkthrough of how to create a book, or experiment yourself.

## Supplements

Adopters of ESource receive an instructor's CD that contains professor and student code from the books in the series, as well as other instruction aides provided by authors. The website also holds approximately **350 PowerPoint transparencies** created by Jack Leifer of University of Kentucky–Paducah available to download. Professors can either follow these transparencies as pre-prepared lectures or use them as the basis for their own custom presentations.

# Titles in the ESource Series

**Introduction to UNIX**

*0-13-095135-8*

*David I. Schwartz*

**Introduction to AutoCAD 2000**

*0-13-016732-0*

*Mark Dix and Paul Riley*

**Introduction to Maple**

*0-13-095133-1*

*David I. Schwartz*

**Introduction to Word**

*0-13-254764-3*

*David C. Kuncicky*

**Introduction to Excel, 2/e**

*0-13-016881-5*

*David C. Kuncicky*

**Introduction to Mathcad**

*0-13-937493-0*

*Ronald W. Larsen*

**Introduction to AutoCAD, R. 14**

*0-13-011001-9*

*Mark Dix and Paul Riley*

**Introduction to the Internet, 3/e**

*0-13-031355-6*

*Scott D. James*

**Engineering Design—A Day in the Life of Four Engineers**

*0-13-085089-6*

*Mark N. Horenstein*

**Engineering Ethics**

*0-13-784224-4*

*Charles B. Fleddermann*

**Mathematics Review**

*0-13-011501-0*

*Peter Schiavone*

**Introduction to C**

*0-13-011854-0*

*Delores Etter*

**Introduction to C++**

*0-13-011855-9*

*Delores Etter*

**Introduction to MATLAB**

*0-13-013149-0*

*Delores Etter with David C. Kuncicky*

**Introduction to FORTRAN 90**

*0-13-013146-6*

*Larry Nyhoff and Sanford Leestma*

**Introduction to Java**

*0-13-919416-9*

*Stephen J. Chapman*

**Introduction to Engineering Analysis**
0-13-016733-9
Kirk D. Hagen

**Introduction to PowerPoint**
0-13-040214-1
Jack Leifer

**Graphics Concepts**
0-13-030687-8
Richard M. Lueptow

**Introduction to MATLAB 6**
0-13-032845-6
Delores Etter and David C. Kuncicky,
with Douglas W. Hull

**Design Concepts for Engineers, 2/e**
0-13-093430-5
Mark Horenstein

**Engineering Success, 2/e**
0-13-041827-7
Peter Schiavone

**Introduction to Engineering Experimentation**
0-13-032835-9
Ronald W. Larsen, John T. Sears, and
Royce Wilkinson

**Graphics Concepts with Pro/ENGINEER**
0-13-014154-2
Richard M. Lueptow, Jim Steger, and
Michael T. Snyder

**Graphics Concepts with SolidWorks**
0-13-014155-0
Richard M. Lueptow and Michael Minbiole

**Introduction to Visual Basic 6.0**
0-13-026813-5
David I. Schneider

**Introduction to Mathcad 2000**
0-13-020007-7
Ronald W. Larsen

**Introduction to Mechanical Engineering**
0-13-019640-1
Robert Rizza

**Introduction to Maple 6**
0-13-032844-8
David I. Schwartz

**Introduction to Electrical and Computer Engineering**
0-13-033363-8
Charles B. Fleddermann and Martin Bradshaw

**Engineering Design and Problem Solving, 2E**
ISBN 0-13-093399-6
Steven K. Howell

**Exploring Engineering**
ISBN 0-13-093442-9
Joe King

# About the Authors

No project could ever come to pass without a group of authors who have the vision and the courage to turn a stack of blank paper into a book. The authors in this series worked diligently to produce their books, provide the building blocks of the series.

**Martin D. Bradshaw** was born in Pittsburg, KS in 1936, grew up in Kansas and the surrounding states of Arkansas and Missouri, graduating from Newton High School, Newton, KS in 1954. He received the B.S.E.E.  and M.S.E.E. degrees from the University of Wichita in 1958 and 1961, respectively. A Ford Foundation fellowship at Carnegie Institute of Technology followed from 1961 to 1963 and he received the Ph.D. degree in electrical engineering in 1964. He spent his entire academic career with the Department of Electrical and Computer Engineering at the University of New Mexico (1961-1963 and 1991-1996). He served as the Assistant Dean for Special Programs with the UNM College of Engineering from 1974 to 1976 and as the Associate Chairman for the EECE Department from 1993 to 1996. During the period 1987-1991 he was a consultant with his own company, EE Problem Solvers. During 1978 he spent a sabbatical year with the State Electricity Commission of Victoria, Melbourne, Australia. From 1979 to 1981 he served an IPA assignment as a Project Officer at the U.S. Air Force Weapons Laboratory, Kirkland AFB, Albuquerque, NM. He has won numerous local, regional, and national teaching awards, including the George Westinghouse Award from the ASEE in 1973. He was awarded the IEEE Centennial Medal in 2000.

*Acknowledgments:* Dr. Bradshaw would like to acknowledge his late mother, who gave him a great love of reading and learning, and his father, who taught him to persist until the job is finished. The encouragement of his wife, Jo, and his six children is a never-ending inspiration.

**Stephen J. Chapman** received a B.S. degree in Electrical Engineering from Louisiana State University (1975), the M.S.E. degree in Electrical Engineering from the University of Central Florida (1979), and pursued further graduate studies at Rice University.

Mr. Chapman is currently Manager of Technical Systems for British Aerospace Australia, in Melbourne, Australia. In this position, he provides technical direction and design authority for the work of younger engineers within the company. He also continues to teach at local universities on a part-time basis.

Mr. Chapman is a Senior Member of the Institute of Electrical and Electronics Engineers (and several of its component societies). He is also a member of the Association for Computing Machinery and the Institution of Engineers (Australia).

**Mark Dix** began working with AutoCAD in 1985 as a programmer for CAD Support Associates, Inc. He helped design a system for creating estimates and bills of material directly from AutoCAD drawing databases for  use in the automated conveyor industry. This system became the basis for systems still widely in use today. In 1986 he began collaborating with Paul Riley to create AutoCAD training materials, combining Riley's background in industrial design and training with Dix's background in writing, curriculum development, and programming. Mr. Dix received the M.S. degree in education from the University of Massachusetts. He is currently the Director of Dearborn Academy High School in Arlington, Massachusetts.

**Delores M. Etter** is a Professor of Electrical and Computer Engineering at the University of Colorado. Dr. Etter was a faculty member at the University of New Mexico and also a Visiting Professor at Stanford University.

Dr. Etter was responsible for the Freshman Engineering Program at the University of New Mexico and is active in the Integrated Teaching Laboratory at the University of Colorado. She was elected a Fellow of the Institute of Electrical and Electronics Engineers for her contributions to education and for her technical leadership in digital signal processing.

**Charles B. Fleddermann** is a professor in the Department of Electrical and Computer Engineering at the University of New Mexico in Albuquerque, New Mexico. All of his degrees are in electrical engineering: his Bachelor's degree from the University of Notre Dame, and the Master's and Ph.D. from the University of Illinois at Urbana-Champaign. Prof. Fleddermann developed an engineering ethics course for his department in response to the ABET requirement to incorporate ethics topics into the undergraduate engineering curriculum. *Engineering Ethics* was written as a vehicle for presenting ethical theory, analysis, and problem solving to engineering undergraduates in a concise and readily accessible way.

*Acknowledgments*: I would like to thank Profs. Charles Harris and Michael Rabins of Texas A & M University whose NSF sponsored workshops on engineering ethics got me started thinking in this field. Special thanks to my wife Liz, who proofread the manuscript for this book, provided many useful suggestions, and who helped me learn how to teach "soft" topics to engineers.

**Kirk D. Hagen** is a professor at Weber State University in Ogden, Utah. He has taught introductory-level engineering courses and upper-division thermal science courses at WSU since 1993. He received his B.S. degree in physics from Weber State College and his M.S. degree in mechanical engineering from Utah State University, after which he worked as a thermal designer/analyst in the aerospace and electronics industries. After several years of engineering practice, he resumed his formal education, earning his Ph.D. in mechanical engineering at the University of Utah. Hagen is the author of an undergraduate heat transfer text.

**Mark N. Horenstein** is a Professor in the Department of Electrical and Computer Engineering at Boston University. He has degrees in Electrical Engineering from M.I.T. and U.C. Berkeley and has been involved in teaching engineering design for the greater part of his academic career. He devised and developed the senior design project class taken by all electrical and computer engineering students at Boston University. In this class, the students work for a virtual engineering company developing products and systems for real-world engineering and social-service clients.

*Acknowledgments:* I would like to thank Prof. James Bethune, the architect of the Peak Performance event at Boston University, for his permission to highlight the competition in my text. Several of the ideas relating to brainstorming and teamwork were derived from a workshop on engineering design offered by Prof. Charles Lovas of Southern Methodist University. The principles of estimation were derived in part from a freshman engineering problem posed by Prof. Thomas Kincaid of Boston University.

**Steven Howell** is the Chairman and a Professor of Mechanical Engineering at Lawrence Technological University. Prior to joining LTU in 2001, Dr. Howell led a knowledge-based engineering project for Visteon Automotive Systems and taught computer-aided design classes for Ford Motor Company engineers. Dr. Howell also has a total of 15 years experience as an engineering faculty member at Northern Arizona University, the University of the Pacific, and the University of Zimbabwe. While at Northern Arizona University, he helped develop and implement an award-winning interdisciplinary series of design courses simulating a corporate engineering-design environment.

**Douglas W. Hull** is a graduate student in the Department of Mechanical Engineering at Carnegie Mellon University in Pittsburgh, Pennsylvania. He is the  author of *Mastering Mechanics I Using Matlab 5*, and contibuted to *Mechanics of Materials* by Bedford and Liechti. His research in the Sensor Based Planning lab involves motion planning for hyper-redundant manipulators, also known as serpentine robots.

**Scott D. James** is a staff lecturer at Kettering University (formerly GMI Engineering & Management Institute) in Flint, Michigan. He is currently pursuing a Ph.D. in Systems Engineering with  an emphasis on software engineering and computer-integrated manufacturing. He chose teaching as a profession after several years in the computer industry. "I thought that it was really important to know what it was like outside of academia. I wanted to provide students with classes that were up to date and provide the information that is really used and needed."

*Acknowledgments*: Scott would like to acknowledge his family for the time to work on the text and his students and peers at Kettering who offered helpful critiques of the materials that eventually became the book.

**Joe King** received the B.S. and M.S. degrees from the University of California at Davis. He is a Professor of Computer Engineering at the University of the Pacific, Stockton, CA, where he teaches  courses in digital design, computer design, artificial intelligence, and computer networking. Since joining the UOP faculty, Professor King has spent yearlong sabbaticals teaching in Zimbabwe, Singapore, and Finland. A licensed engineer in the state of California, King's industrial experience includes major design projects with Lawrence Livermore National Laboratory, as well as independent consulting projects. Prof. King has had a number of books published with titles including MATLAB, MathCAD, Exploring Engineering, and Engineering and Society.

**David C. Kuncicky** is a native Floridian. He earned his Baccalaureate in psychology, Master's in computer science, and Ph.D. in computer science from  Florida State University. He has served as a faculty member in the Department of Electrical Engineering at the FAMU–FSU College of Engineering and the Department of Computer Science at Florida State University. He has taught computer science and computer engineering courses for over 15 years. He has published research in the areas of intelligent hybrid systems and neural networks. He is currently the Director of Engineering at Bioreason, Inc. in Sante Fe, New Mexico.

*Acknowledgments:* Thanks to Steffie and Helen for putting up with my late nights and long weekends at the computer. Finally, thanks to Susan Bassett for having faith in my abilities, and for providing continued tutelage and support.

**Ron Larsen** is a Professor of Chemical Engineering at Montana State University, and received his Ph.D. from the Pennsylvania State University. He was initially attracted to engineering by the challenges the profession offers, but also  appreciates that engineering is a serving profession. Some of the greatest challenges he has faced while teaching have involved non-traditional teaching methods, including evening courses for practicing engineers and teaching through an interpreter at the Mongolian National University. These experiences have provided tremendous opportunities to learn new ways to communicate technical material. Dr. Larsen views modern software as one of the new tools that will radically alter the way engineers work, and his book *Introduction to MathCAD* was written to help young engineers prepare to meet the challenges of an ever-changing workplace.

*Acknowledgments*: To my students at Montana State University who have endured the rough drafts and typos, and who still allow me to experiment with their classes— my sincere thanks.

**Sanford Leestma** is a Professor of Mathematics and Computer Science at Calvin College, and received his Ph.D. from New Mexico State University. He has been the long-time co-author of successful textbooks on Fortran, Pascal, and data structures in Pascal. His current research interest are in the areas of algorithms and numerical computation.

**Jack Leifer** is an Assistant Professor in the Department of Mechanical Engineering at the University of Kentucky Extended Campus Program in Paducah, and was previously with the Department of Mathematical Sciences and Engineering at the University of South Carolina–Aiken. He received his Ph.D. in Mechanical Engineering from the University of Texas at Austin in December 1995. His current research interests include the modeling of sensors for manufacturing, and the use of Artificial Neural Networks to predict corrosion.

*Acknowledgements:* I'd like to thank my colleagues at USC–Aiken, especially Professors Mike May and Laurene Fausett, for their encouragement and feedback; and my parents, Felice and Morton Leifer, for being there and providing support (as always) as I completed this book.

**Richard M. Lueptow** is the Charles Deering McCormick Professor of Teaching Excellence and Associate Professor of Mechanical Engineering at Northwestern University. He is a native of Wisconsin and received his doctorate from the Massachusetts Institute of Technology in 1986. He teaches design, fluid mechanics, and spectral analysis techniques. Rich has an active research program on rotating filtration, Taylor Couette flow, granular flow, fire suppression, and acoustics. He has five patents and over 40 refereed journal and proceedings papers along with many other articles, abstracts, and presentations.

*Acknowledgments:* Thanks to my talented and hard-working co-authors as well as the many colleagues and students who took the tutorial for a "test drive." Special thanks to Mike Minbiole for his major contributions to Graphics Concepts with SolidWorks. Thanks also to Northwestern University for the time to work on a book. Most of all, thanks to my loving wife, Maiya, and my children, Hannah and Kyle, for supporting me in this endeavor. (Photo courtesy of Evanston Photographic Studios, Inc.)

**Larry Nyhoff** is a Professor of Mathematics and Computer Science at Calvin College. After doing bachelor's work at Calvin, and Master's work at Michigan, he received a Ph.D. from Michigan State and also did graduate work in computer science at Western Michigan. Dr. Nyhoff has taught at Calvin for the past 34 years—mathematics at first and computer science for the past several years.

*Acknowledgments*: We thank our families—Shar, Jeff, Dawn, Rebecca, Megan, Sara, Greg, Julie, Joshua, Derek, Tom, Joan; Marge, Michelle, Sandy, Lory, Michael—for being patient and understanding. We thank God for allowing us to write this text.

**Paul Riley** is an author, instructor, and designer specializing in graphics and design for multimedia. He is a founding partner of CAD Support Associates, a contract service and professional training organization for computer-aided design. His 15 years of business experience and 20 years of teaching experience are supported by degrees in education and computer science. Paul has taught

AutoCAD at the University of Massachusetts at Lowell and is presently teaching AutoCAD at Mt. Ida College in Newton, Massachusetts. He has developed a program, Computer-aided Design for Professionals that is highly regarded by corporate clients and has been an ongoing success since 1982.

**Robert Rizza** is an Assistant Professor of Mechanical Engineering at North Dakota State University, where he teaches courses in mechanics and computer-aided design. A native of Chicago, he received the Ph.D. degree from the Illinois Institute of Technology. He is also the author of *Getting Started with Pro/ENGINEER*. Dr. Rizza has worked on a diverse range of engineering projects including projects from the railroad, bioengineering, and aerospace industries. His current research interests include the fracture of composite materials, repair of cracked aircraft components, and loosening of prostheses.

**Peter Schiavone** is a professor and student advisor in the Department of Mechanical Engineering at the University of Alberta, Canada. He received his Ph.D. from the University of Strathclyde, U.K. in 1988. He has authored several books in the area of student academic success as well as numerous papers in international scientific research journals. Dr. Schiavone has worked in private industry in several different areas of engineering including aerospace and systems engineering. He founded the first Mathematics Resource Center at the University of Alberta, a unit designed specifically to teach new students the necessary *survival skills* in mathematics and the physical sciences required for success in first-year engineering. This led to the Students' Union Gold Key Award for outstanding contributions to the university. Dr. Schiavone lectures regularly to freshman engineering students and to new engineering professors on engineering success, in particular about maximizing students' academic performance.

*Acknowledgements:* Thanks to Richard Felder for being such an inspiration; to my wife Linda for sharing my dreams and believing in me; and to Francesca and Antonio for putting up with Dad when working on the text.

**David I. Schneider** holds an A.B. degree from Oberlin College and a Ph.D. degree in Mathematics from MIT. He has taught for 34 years, primarily at the University of Maryland. Dr. Schneider has authored 28 books, with one-half of them computer programming books. He has developed three customized software packages that are supplied as supplements to over 55 mathematics textbooks. His involvement with computers dates back to 1962, when he programmed a special purpose computer at MIT's Lincoln Laboratory to correct errors in a communications system.

**David I. Schwartz** is an Assistant Professor in the Computer Science Department at Cornell University and earned his B.S., M.S., and Ph.D. degrees in Civil Engineering from State University of New York at Buffalo. Throughout his graduate studies, Schwartz combined principles of computer science to applications of civil engineering. He became interested in helping students learn how to apply software tools for solving a variety of engineering problems. He teaches his students to learn incrementally and practice frequently to gain the maturity to tackle other subjects. In his spare time, Schwartz plays drums in a variety of bands.

*Acknowledgments:* I dedicate my books to my family, friends, and students who all helped in so many ways. Many thanks go to the schools of Civil Engineering and Engineering & Applied Science at State University of New York at Buffalo where I originally developed and tested my UNIX and Maple books. I greatly appreciate the opportunity to explore my goals and all the help from everyone at the Computer Science Department at Cornell.

**John T. Sears** received the Ph.D. degree from Princeton University. Currently, he is a Professor and the head of the Department of Chemical Engineering at Montana State University. After leaving Princeton he worked in research at Brookhaven National Laboratory and Esso Research and Engineering, until he took a position at West Virginia University. He came to MSU in 1982, where he has served as the Director of the College of Engineering Minority Program and Interim Director for BioFilm Engineering. Prof. Sears has written a book on air pollution and economic development, and over 45 articles in engineering and engineering education.

**Michael T. Snyder** is President of Internet startup Appointments123.com. He is a native of Chicago, and he received his Bachelor of Science degree in Mechanical Engineering from the University of Notre Dame. Mike also graduated with honors from Northwestern University's Kellogg Graduate School of Management in 1999 with his Masters of Management degree. Before Appointments123.com, Mike was a mechanical engineer in new product development for Motorola Cellular and Acco Office Products. He has received four patents for his mechanical design work. "Pro/ENGINEER was an invaluable design tool for me, and I am glad to help students learn the basics of Pro/ENGINEER."

*Acknowledgments:* Thanks to Rich Lueptow and Jim Steger for inviting me to be a part of this great project. Of course, thanks to my wife Gretchen for her support in my various projects.

**Jim Steger** is currently Chief Technical Officer and cofounder of an Internet applications company. He graduated with a Bachelor of Science degree in Mechanical Engineering from Northwestern University. His prior work included mechanical engineering assignments at Motorola and Acco Brands. At Motorola, Jim worked on part design for two-way radios and was one of the lead mechanical engineers on a cellular phone product line. At Acco Brands, Jim was the sole engineer on numerous office product designs. His Worx stapler has won design awards in the United States and in Europe. Jim has been a Pro/ENGINEER user for over six years.

*Acknowledgments:* Many thanks to my co-authors, especially Rich Lueptow for his leadership on this project. I would also like to thank my family for their continuous support.

**Royce Wilkinson** received his undergraduate degree in chemistry from Rose-Hulman Institute of Technology in 1991 and the Ph.D. degree in chemistry from Montana State University in 1998 with research in natural product isolation from fungi. He currently resides in Bozeman, MT and is involved in HIV drug research. His research interests center on biological molecules and their interactions in the search for pharmaceutical advances.

# Reviewers

ESource benefited from a wealth of reviewers who on the series from its initial idea stage to its completion. Reviewers read manuscripts and contributed insightful comments that helped the authors write great books. We would like to thank everyone who helped us with this project.

## Concept Document

Naeem Abdurrahman  *University of Texas, Austin*
Grant Baker  *University of Alaska, Anchorage*
Betty Barr  *University of Houston*
William Beckwith  *Clemson University*
Ramzi Bualuan  *University of Notre Dame*
Dale Calkins  *University of Washington*
Arthur Clausing  *University of Illinois at Urbana–Champaign*
John Glover  *University of Houston*
A.S. Hodel  *Auburn University*
Denise Jackson  *University of Tennessee, Knoxville*
Kathleen Kitto  *Western Washington University*
Terry Kohutek  *Texas A&M University*
Larry Richards  *University of Virginia*
Avi Singhal  *Arizona State University*
Joseph Wujek  *University of California, Berkeley*
Mandochehr Zoghi  *University of Dayton*

## Books

Stephen Allan  *Utah State University*
Naeem Abdurrahman  *University of Texas, Austin*
Anil Bajaj  *Purdue University*
Grant Baker  *University of Alaska–Anchorage*
Betty Burr  *University of Houston*
William Beckwith  *Clemson University*
Haym Benaroya  *Rutgers University*
Tom Bledsaw  *ITT Technical Institute*
Tom Bryson  *University of Missouri, Rolla*
Ramzi Bualuan  *University of Notre Dame*
Dan Budny  *Purdue University*
Dale Calkins  *University of Washington*
Arthur Clausing  *University of Illinois*
James Devine  *University of South Florida*

Patrick Fitzhorn  *Colorado State University*
Dale Elifrits  *University of Missouri, Rolla*
Frank Gerlitz  *Washtenaw College*
John Glover  *University of Houston*
John Graham  *University of North Carolina–Charlotte*
Malcom Heimer  *Florida International University*
A.S. Hodel  *Auburn University*
Vern Johnson  *University of Arizona*
Kathleen Kitto  *Western Washington University*
Robert Montgomery  *Purdue University*
Mark Nagurka  *Marquette University*
Romarathnam Narasimhan  *University of Miami*
Larry Richards  *University of Virginia*
Marc H. Richman  *Brown University*
Avi Singhal  *Arizona State University*
Tim Sykes  *Houston Community College*
Thomas Hill  *SUNY at Buffalo*
Michael S. Wells  *Tennessee Tech University*
Joseph Wujek  *University of California, Berkeley*
Edward Young  *University of South Carolina*
Mandochehr Zoghi  *University of Dayton*
John Biddle  *California State Polytechnic University*
Fred Boadu  *Duke University*
Harish Cherukuri  *University of North Carolina–Charlotte*
Barry Crittendon  *Virginia Polytechnic and State University*
Ron Eaglin  *University of Central Florida*
Susan Freeman  *Northeastern University*
Frank Gerlitz  *Washtenaw Community College*
Otto Gygax  *Oregon State University*
Donald Herling  *Oregon State University*
James N. Jensen  *SUNY at Buffalo*
Autar Kaw  *University of South Florida*
Kenneth Klika  *University of Akron*
Terry L. Kohutek  *Texas A&M University*
Melvin J. Maron  *University of Louisville*
Soronadi Nnaji  *Florida A&M University*
Michael Peshkin  *Northwestern University*
Randy Shih  *Oregon Institute of Technology*
Neil R. Thompson  *University of Waterloo*
Garry Young  *Oklahoma State University*

# Contents

**1  WHAT IS ENGINEERING?  1**

1.1 Engineering Has Many Fields 2
 1.1.1 Aeronautical Engineer 2
 1.1.2 Agricultural Engineer 3
 1.1.3 Biomedical Engineer 3
 1.1.4 Chemical Engineer 4
 1.1.5 Civil Engineer 4
 1.1.6 Computer Engineer 6
 1.1.7 Electrical Engineer 6
 1.1.8 Industrial Engineer 6
 1.1.9 Mechanical Engineer 8
 1.1.10 Mechatronics Engineer 9
 1.1.11 Naval Engineer 9
 1.1.12 Petroleum Engineer 9
 1.1.13 Systems Engineer 10
1.2 Professional Engineering Organizations 10
 1.2.1 Aeronautical and Aerospace Engineering 11
 1.2.2 Biomedical Engineering 11
 1.2.3 Chemical Engineering 12
 1.2.4 Civil Engineering 12
 1.2.5 Computer Engineering 13
 1.2.6 Electrical Engineering 13
 1.2.7 Industrial and Manufacturing Engineering 14
 1.2.8 Mechanical Engineering 14
1.3 The Engineer: Central to Project Management 15
 1.3.1 The Well-Rounded Engineer 16
1.4 Engineering: A Set of Skills 19
 1.4.1 Knowledge 19
 1.4.2 Experience 20
 1.4.3 Intuition 20

**2  WHAT IS DESIGN?  22**

2.1 The Use of the Word "Design" 23
2.2 The Difference Between Design, Analysis, and Reproduction 23
2.3 Good Design Versus Bad Design 29
2.4 The Design Cycle 33
 2.4.1 Define the Overall Objectives 33
 2.4.2 Gather Information 33
 2.4.3 Choose a Design Strategy 33
 2.4.4 Make a First Cut at the Design 35
 2.4.5 Build, Document, and Test 35
 2.4.6 Revise and Revise Again 36
 2.4.7 Thoroughly Test the Finished Product 37
2.5 A Design Example 38
 2.5.1 Applying Design Principles to the Design Competition 40

2.5.2    Define the Overall Objectives   40
2.5.3    Choose a Design Strategy   41
2.5.4    Make a First Cut at the Design   44
2.5.5    Build, Document, Test, and Revise   44
2.5.6    Revise Again   45
2.5.7    Reality Check   46
2.5.8    More Revisions   46

# 3   WORKING IN TEAMS   60

3.1    Teamwork Skills   61
    3.1.1    Effective Team Building   61
3.2    Brainstorming   63
    3.2.1    Ground Rules for Brainstorming   64
    3.2.2    Formal Brainstorming Method   64
    3.2.3    Informal Brainstorming   69
3.3    Documentation: The Key to Project Success   74
    3.3.1    Paper Versus Electronic Documentation   75
    3.3.2    The Engineers' Logbook   75
    3.3.3    Logbook Format   76
    3.3.4    Using Your Engineer's Logbook   77
    3.3.5    Technical Reports and Memoranda   79
    3.3.6    Schematics and Drawings   79
    3.3.7    Software Documentation and the Role of the Engineering Notebook   79
    3.3.8    The Importance of Logbooks: A Case Study   81
3.4    Project Management: Keeping the Team on Track   83
    3.4.1    Organizational Chart   83
    3.4.2    Time Line   84
    3.4.3    Gantt Chart   84

# 4   ENGINEERING DESIGN TOOLS   90

4.1    Estimation   91
4.2    Significant Figures, Dimensioning, and Tolerance   96
4.3    Prototyping and Breadboarding   98
4.4    Reverse Engineering   107
4.5    Computer Analyses   107
    4.5.1    Plotting Trajectories Using MATLAB   112
4.6    The Internet   120
    4.6.1    "I Saw It on the Internet. It Must be True."   120
4.7    Spreadsheets in Engineering Design   121
4.8    Solid Modeling and Computer-Aided Drafting   128
    4.8.1    Why an Engineering Drawing?   129
    4.8.2    Types of Drawings   130
4.9    System Simulation   135
4.10    Electronic Circuit Simulation   136
4.11    Graphical Programming   139
4.12    Microprocessors: The "Other" Computer   140
    4.12.1    Analog-to-Digital and Digital-to-Analog Conversion   141

# 5   THE HUMAN–MACHINE INTERFACE   155

5.1    How People Interact with Machines   156
5.2    Ergonomics   156
    5.2.1    Putting Ergonomics to Work   157
5.3    Cognition   158
5.4    The Human–Machine Interface: Case Studies   159

**6**  ENGINEERS AND THE REAL WORLD   178

6.1   Society's View of Engineering   178
6.2   How Engineers Learn From Mistakes   181
6.3   The Role of Failure in Engineering Design: Case Studies   182
      6.3.1   Case 1: Tacoma Narrows Bridge   183
      6.3.2   Case 2: Hartford Civic Center   183
      6.3.3   Case 3: Space Shuttle *Challenger*   185
      6.3.4   Case 4: Kansas City Hyatt   187
      6.3.5   Case 5: Three Mile Island   189
      6.3.6   Case 6: USS Vincennes   191
      6.3.7   Case 7: Hubble Telescope   192
      6.3.8   Case 8: De Haviland Comet   192
6.4   Preparing for Failure in Your Own Design   193

**7**  LEARNING TO SPEAK, WRITE, AND MAKE PRESENTATIONS   195

7.1   The Importance of Good Communication Skills   196
7.2   Preparing for Meetings, Presentations, and Conferences   196
7.3   Preparing for a Formal Presentation   198
7.4   Writing Electronic Mail, Letters, and Memoranda   203
      7.4.1   Writing Electronic Mail Messages   204
      7.4.2   Header   204
      7.4.3   First Sentence   205
      7.4.4   Body   206
      7.4.5   Writing Formal Memos and Letters   208
7.5   Writing Technical Reports, Proposals, and Journal Articles   212
      7.5.1   Technical Report   212
      7.5.2   Journal Paper   213
      7.5.3   Proposal   213
7.6   Preparing an Instruction Manual   213
      7.6.1   Introduction   213
      7.6.2   Setup   214
      7.6.3   Operation   214
      7.6.4   Safety   214
      7.6.5   Troubleshooting   214
      7.6.6   Appendices   215
      7.6.7   Repetition   215
7.7   Producing Good Technical Documents: A Strategy   219
      7.7.1   Plan the Writing Task   219
      7.7.2   Find a Place to Work   219
      7.7.3   Define the Reader   220
      7.7.4   Make Notes   220
      7.7.5   Create Topic Headings   220
      7.7.6   Take a Break   220
      7.7.7   Write the First Draft   220
      7.7.8   Read the Draft   221
      7.7.9   Revise the Draft   221
      7.7.10  Revise, Revise, and Revise Again   221
      7.7.11  Review the Final Draft   222
      7.7.12  Common Writing Errors   222

**8**  THE DAY OF THE PEAK PERFORMANCE DESIGN COMPETITION   227

INDEX   232

# 1

# What Is Engineering?

If you're reading this book, you're probably enrolled in an introductory course in *engineering*. You may have chosen engineering because of your strong skills in science and mathematics. Perhaps you like to take things apart or use computers. Maybe you simply followed the advice of your high school guidance counselor. Whatever your reason for studying engineering, you are entering a career full of discovery, creativity, and excitement. Imagine yourself several years from now, after you've finished your college studies. What will your life be like as an engineer? How will the classes you've taken in school relate to your work and career? This book will help provide you with a vision of the future while teaching you the important principles of engineering design.

As an aspiring engineer, you have much to learn. You must master the foundations of all engineering disciplines: basic math, physics, and chemistry. You must study the specialized subjects of your chosen discipline, for example, circuits, mechanics, materials, or computer programming. You also must learn how to stay on top of technological advances by embracing a program of lifelong learning. The world embraces new technological advances almost on a daily basis, and the wise engineer keeps abreast of them all. Your

## SECTIONS

- 1.1 Engineering Has Many Fields
- 1.2 Professional Engineering Organizations
- 1.3 The Engineer: Central to Project Management
- 1.4 Engineering: A Set of Skills

## OBJECTIVES

*In this chapter, you will learn about*

- Engineering as a career.
- The relationship between the engineer and other professionals.
- Engineering professional organizations.
- The foundations of engineering design: knowledge, experience, and intuition.

college courses will provide you with the knowledge and mathematical skills that you will need to function in the engineering world. However, you also must learn about the primary mission of the engineer: the practice of design. The ability to build real things is what sets an engineer apart from professionals in the basic sciences. While physicists, chemists, and biologists draw general conclusions by observing specific phenomena, an engineer moves from the general to the specific. Engineers harness the laws of nature and use them to produce devices or systems that perform tasks and solve problems. This process defines the essence of design. You must become proficient at it if you want to become an engineer. This book will teach you the principles of design and help you to apply them to your class assignments, design projects, and future job activities.

## 1.1 ENGINEERING HAS MANY FIELDS

A perusal of the Web sites of engineering colleges around the world will reveal a wide variety of engineering programs. Although the names may vary slightly, most engineers are trained in one of the following traditional engineering degree programs (listed alphabetically): aeronautical, agricultural, biomedical, chemical, civil, computer, electrical, industrial, mechanical, naval, petroleum, and systems. From reading this list, one might get the impression that engineers are highly specialized professionals who have little interaction with people from other fields. In reality, the opposite is true. The best engineers are multidisciplinary individuals who are familiar with many different fields and specialties. The mechanical engineer knows something about electrical circuits, and the electrical engineer understands basic mechanics. The computer engineer is familiar with the algorithms used in industrial processes, and the industrial engineer knows how to program computers. Many of the great engineering accomplishments of the past century, including our global communication network; the Internet; life-extending biomedical technology; inexpensive, reliable air transportation; our ground transportation infrastructure; and the sequencing of the human genome were made possible by interactive teams of engineers from many disciplines.

Although engineers have multidisciplinary skills, most are trained in a specific degree program and spend much time using their specialized training. For this reason, we precede our study of design by reviewing the characteristic features of the various types of engineers and their fields of expertise.

### 1.1.1 Aeronautical Engineer

Aeronautical (or aerospace) engineers use their knowledge of aerodynamics, fluid mechanics, structures, control systems, heat transfer, and hydraulics to design and build everything from rockets, airplanes, and space vehicles to high-speed bullet trains and helium-filled dirigibles. Since the days of the Wright brothers, aeronautical engineers, working in teams with scientists and other types of engineers, have made possible human flight and space exploration. Aeronautical engineers find employment in many industries, but typically work for big companies on large-scale projects. Some of the more noticeable accomplishments of the aerospace industry have included the Apollo moon landings, the NASA Space Shuttle, deep space exploration, space stations, and the jumbo jet. The International Space Station, shown in Figure 1.1, for example, will be completed by teams of aeronautical and other engineers.

**Figure 1.1.**    Artist's view of NASA's International Space Station. This station is being built by teams of aerospace and aeronautical engineers working together with other types of engineers and scientists. (*Photo courtesy of NASA.*)

### 1.1.2    Agricultural Engineer

Agricultural engineers apply the principles of hydrology, soil mechanics, fluid mechanics, heat transfer, combustion, optimization theory, statistics, climatology, chemistry, and biology to the production of food on a large scale. This discipline is popular at colleges and universities located in heavily agricultural areas. Feeding the world's ever-growing population is one of the most formidable challenges of the 21st century. Agricultural engineers will play an important role in this endeavor by applying technology and engineering know-how to improve crop yields, increase food output, and develop cost-effective and environmentally sound farming methods. Agricultural engineers work with ecologists and biologists, chemists, and natural scientists to understand the impact of human agriculture on the earth's ecosystem.

### 1.1.3    Biomedical Engineer

The biomedical engineer (or bioengineer) works closely with physicians and biologists to apply modern engineering methods to medicine and human health and to obtain a better understanding of the human body. Engineering skills are combined with knowledge of biology, physiology, and chemistry to produce medical instrumentation, prosthetics, assistive appliances, implants, and neuromuscular diagnostics. Biomedical engineers have participated in designing many devices that have helped improve medical care over the past several decades. Many biomedical engineers enter medical school upon graduation, but others go on to graduate school or seek employment in any of a number of health- or medical-related industries. The rapidly emerging world of biotechnology, which bridges the gap between engineering and molecular genetics, is also the province of the biomedical engineer (Figure 1.2). This discipline examines the fundamental functions of cells and organisms from an engineering point of view. The science of cloning, for example, has been made possible in part by the world of molecular engineers. Many of the secrets of future medicine lie at the genetic level, and the biomedical engineer will help lead the way to new medical discoveries. The biomedical engineer also is involved in the area of microfluidics, in which tiny bio-processing systems are built on small chips of silicon or other materials. This technology, part of the field of micro-electromechanical systems, or MEMS, is sometimes referred to as "lab on a chip." Another exciting area of biomedical engineering is the field of *bioinformatics* which combines computer science with the study of genomics, including the human genome.

### 1.1.4   Chemical Engineer

The chemical engineer applies the principles of chemistry to the design of manufacturing and production systems. Whenever a chemical reaction or process must be brought from the laboratory to manufacturing on a large scale, a chemical engineer is needed to design the reaction vessels, transport mechanisms, mixing chambers, and measuring devices that allow the process to proceed on a large scale in a cost-effective way. Chemical engineers are employed in many industries, including petroleum, petrochemicals, plastics, cosmetics, electronics, food, and pharmaceuticals. Their skills are needed wherever a manufacturing process involves organic or inorganic chemical reactions on a production scale. Typically, chemical engineers are employed by large companies that produce products for worldwide distribution.

### 1.1.5   Civil Engineer

The civil engineer is concerned with the design and construction of our nation's infrastructure. Civil engineers design transportation systems, roads, bridges, buildings, airports, and other large structures, such as water treatment plants, aquifers, and waste management facilities. One classic example of civil engineering on a grand scale is the Hoover Dam shown in Figure 1.3. Designing such a large structure requires knowledge of fluid and soil mechanics, strength of materials, concrete engineering, and construction practices. Civil engineers also may be involved in designing smaller structures such as houses, landscapes, and recreational parks. Over the next few decades, civil engineers will play a vital role in revitalizing aging infrastructures worldwide and in dealing with environmental issues, such as water resources, air quality, global warming, and refuse disposal. More than any other professional, the civil engineer has the unique handicap of having to rely heavily on physical scale models, calculations, computer modeling, and past experience to determine the performance of designed structures. This limitation exists because it is seldom possible to build a trial test structure on the scale of most civil engineering products. There are no full-scale prototypes in civil engineering. A civil engineer must be sure that a design will meet its specifications well before its final construction.

The civil engineer works closely with construction personnel and may spend much time at job sites reviewing the progress of construction tasks. Civil engineers often are employed in the public sector, but also may find work in large or small construction companies and private development firms. One renowned example of a large, public-sector civil engineering effort is the famed "Big Dig" in Boston, Massachusetts (Figure 1.4). This multibillion-dollar, 10-year effort, the most expensive and extensive single transportation infrastructure project in U.S. history, is formally known as Central Artery/Tunnel Project.

**Figure 1.2.** Genetics is one of the new frontiers of biomedical engineering. (*Graphic courtesy of the Center for Advanced Biotechnology, Boston University.*)

**Figure 1.3.** The Hoover Dam at the Nevada–Arizona border was designed by civil engineers.

**Figure 1.4.** Boston's Central Artery/Tunnel Project: The "Big Dig." Civil engineers will be responsible for revitalizing the nation's infrastructure in the 21st century. (*Photo courtesy of the Central Artery/Tunnel Project.*)

### 1.1.6 Computer Engineer

Computer engineering encompasses the broad categories of hardware, software, and digital communication (Figure 1.5). A computer engineer applies the basic principles of engineering and computer science to the design of computers, networks, software systems, and peripheral devices. The computer engineer also is responsible for designing and building the interconnections between computers and their components, including distributed computers, local area networks (LANs), wireless networks, and Internet servers. For example, a computer engineer might combine microprocessors, memory chips, disk drives, DVD drives, display devices, LAN cards, and drivers to produce computer systems. Graphical user interfaces, embedded computer systems, fault-tolerant computers, software systems, wireless interfaces, operating systems, and assembly language programming are also the responsibility of the computer engineer. Computer scientists, who traditionally are more mathematically oriented than are computer engineers, also become involved in writing computer software, including Web interfaces, database management systems, and client applications. Unlike the computer scientist, however, the computer engineer is fluent in both the hardware and software aspects of modern computer systems. Examples of which both hardware and software share equally important roles include CPU design, desktop and laptop PC design, cell-phone networks, global positioning systems, microcomputer- or Internet-controlled appliances, automated manufacturing, and medical instrumentation. Some of the more notable accomplishments of the computer industry include the invention of the microprocessor (Intel, 1982), the explosion of personal computing that began with the first desktop PC (IBM, 1984), and the advances in data communication networks that began with the U.S. Department of Defense Arpanet and grew into the Internet and the World Wide Web.

### 1.1.7 Electrical Engineer

Electrical engineering is a far-reaching discipline whose subjects are linked by a single common thread: the use and control of electricity. Because information can be expressed in electronic form, computers also fall into this broad category. Whether on a large or small scale, electrical engineers are responsible for many technology areas including microelectronics, data communication, radio, television, lasers (Figure 1.6), fiber optics, video, audio, computer networks, speech processing, imaging systems, electric power systems, and alternative energy sources, such as solar and wind power. The electrical engineer also designs transportation systems based on electric power, including mass transit, electric cars, and hybrid vehicles.

The typical electrical engineer has a strong background in the physical sciences, mathematics, and computational methods, as well as knowledge of circuits and electronics, semiconductor devices, analog and digital signal processing, digital systems, electromagnetics, and control systems. The electrical engineer also is fluent in many areas of computer engineering. Some of the more recent accomplishments that have involved electrical engineers include the microelectronic revolution (e.g., microprocessors and large-scale integration on a chip); wireless communications (e.g., cellular telephones, pagers, and data links); photonics (e.g., lightwave technology, lasers, and fiber-optic communication); and micro-electromechanical systems (e.g., laboratory on a chip).

### 1.1.8 Industrial Engineer

The industrial engineer (sometimes called a manufacturing engineer) is concerned with the total life cycle of a product, from the moment of its inception to its eventual disposal and recycling of its raw materials. Industrial engineers have the unique challenge of incorporating the latest technological advances in computing and machinery into production and manufacturing facilities. The industrial engineer is intimate with all aspects of the corporate environment, because much of what motivates the field of industrial engineering is the need to maximize output while minimizing cost. Skills required for this discipline include knowledge of product development, materials processing, optimization, queuing theory, production techniques, machining, fabrication methods, and engineering economy. Industrial engineers also become fluent in the techniques of computer-aided design (CAD) and computer-aided manufacturing (CAM). Global manufacturing, in which products are developed for a worldwide economy, is becoming increasingly important to the field of industrial engineering.

One of the more recent areas to emerge as the province of the industrial engineer is the use of robotics in manufacturing. Building, moving, and controlling robots requires knowledge of aspects of mechanical, electrical, and computer engineering. Most programs in industrial engineering include courses in these other areas. Another emerging area of industrial engineering is the field of "green manufacturing" in which an understanding of the environmental impact of a product over its life cycle is considered as part of the design process.

***Figure 1.5.*** Computer engineers design the hardware and software for today's computer systems. (*Image courtesy of C. Moreira and L. Katz.*)

***Figure 1.6.*** The laser, first invented around 1960, has become an indispensable tool for the electrical engineer. (*Photo courtesy of Boston University Photo Services.*)

### 1.1.9 Mechanical Engineer

The mechanical engineer is responsible for designing and building physical structures of all sorts. Devices that involve mechanical motion, such as automobiles, bicycles, engines, disk drives, keyboards, fluid valves, jet engine turbines, power plants, and flight structures, are all designed by mechanical engineers. Mechanical engineers are fluent in the topics of statics, dynamics, strength of materials, structural and solid mechanics, fluid mechanics, thermodynamics, heat transfer, and energy conversion. They apply these principles to a wide variety of engineering problems, including acoustics, precision machining, environmental engineering, water resources, combustion, power sources, robotics, transportation, and manufacturing systems. Compared with all engineering disciplines, the mechanical engineer interfaces most easily with other types of engineers because mechanical engineering requires such a broad educational background.

One of the newest areas of study involving mechanical engineers is the emerging field of micro-electromechanical systems, or MEMS, in which tiny microscopic machines are fabricated on wafers of silicon and other materials. Figure 1.7, for example, shows a tiny micromotor built on a silicon chip. MEMS has the potential to do for mechanics what the integrated circuit did for electronics, namely permit large-scale integration on single silicon chips of entire systems made from basic devices. Mechanical engineers work closely with electrical engineers in this exciting new discipline.

### 1.1.10 Mechatronics Engineer

The mechatronics engineer is fluent in mechanical engineering, electrical engineering, and robotics. As its name implies, the field of mechatronics involves the fusion of mechanical engineering, electronics, and computing toward the design of products and manufacturing systems. Engineers who work in this emerging field require cross-disciplinary training that can best be approached by majoring in either mechanical or electrical engineering and acquiring skills in the other needed disciplines through extra courses or technical electives. Mechatronics engineers are responsible for the innovation, design, and development of machines and systems that can automate production tasks, reduce production costs, reduce plant maintenance costs, improve product flexibility, and increase production performance. The typical mechatronics engineer solves design problems for which solely mechanical or electrical solutions are not possible. Sensing and actuation are important elements of mechatronics.

### 1.1.11 Naval Engineer

The naval engineer (or naval architect) designs ships, submarines, barges, and other sea-going vessels and also is involved in the design of oil platforms, shipping docks, seaports, and coastal navigation facilities. Naval engineers are fluent in many of the subjects studied by mechanical engineers, including fluid mechanics, materials, structures, statics, dynamics, water propulsion, and heat transfer. In addition, naval engineers learn about the design of ships and the history of sea travel. Many naval engineers are employed by the armed forces, but some work for companies that design and build large ships.

### 1.1.12 Petroleum Engineer

Over 70 percent of the world's current energy needs are satisfied by petroleum products, and this situation is unlikely to change for at least the next half century. The principle challenge of the petroleum engineer is to help produce oil, gas, and other energy forms from the earth's natural resources. In order to harvest these resources in an economical and environmentally safe way, the petroleum engineer must have a wide base of knowledge that includes mathematics, physics, geology, and chemistry, as well as aspects of most other engineering disciplines. Elements of mechanical, chemical, electrical, civil, and industrial engineering are found in most programs of study in petroleum engineering. Also, because computers are used with ever-increasing frequency in geological exploration, oil field production, and drilling operations, computer engineering has become an important specialty within petroleum engineering. Did you know that many of the world's supercomputers are owned by petroleum companies?

**Figure 1.7.**   Tiny MEMS silicon micromotor measuring 100 micrometers in diameter. (*Photo courtesy of Cronos, Inc.*)

In addition to conventional oil and gas recovery, petroleum engineers apply new technology to the enhanced recovery of hydrocarbons from oil shale, tar sands, offshore oil deposits, and fields of natural gas. They also design new techniques for recovering residual ground oil that has been left by traditional pumping methods. Examples include the use of underground combustion, steam injection, and chemical water treatment to release oil trapped in the pores of rock. These techniques will likely be used in the future for other geological operations, including uranium leaching, geothermal energy production, and coal gasification. Petroleum engineers also work in the related areas of pollution reduction, underground waste disposal, and hydrology. Lastly, because many petroleum companies operate on a worldwide scale, the petroleum engineer has the opportunity to work in many foreign countries.

### 1.1.13  Systems Engineer

In the computer industry, the designation "systems engineer" has come to mean someone who deals exclusively with large-scale software systems. The traditional systems engineer, however, can be anyone who designs and implements complex engineering systems. Communications systems, transportation networks, manufacturing systems, power distribution networks, and avionics are but some of the areas in which systems engineers play a central role. Programs of study in this diverse field include courses in applied mathematics, computer simulation, software, electronics, communications, and automatic control. Because of their broad educational background, systems engineers are at home working with most other types of engineers.

## 1.2   PROFESSIONAL ENGINEERING ORGANIZATIONS

Most branches of engineering are represented by professional societies that bind together members with similar backgrounds, training, and professional expertise. These societies operate on a worldwide scale and publish one or more journals for which engineers write papers and articles of interest to members of the field. Each organization offers its members technical and informational services, including training, industry standards, workshops, and conferences. In some cases, other professional services are offered as well, including job networks, advertising, e-mail accounts, product information, Web page hosting, and even life and health insurance. All provide student membership at a discount, and student chapters at colleges and universities are common. This section provides information about some of the principal professional organizations and the technical publications they produce. Each society has an official Web site from which you can obtain additional information. The text provided here has been taken from each organization's Web site.

### 1.2.1  Aeronautical and Aerospace Engineering

From the American Institute of Aeronautics and Astronautics (*www.aiaa.org*):

> "For more than 65 years, the American Institute of Aeronautics and Astronautics (AIAA) and its predecessors, has been the principal society of the aerospace engineer and scientist. Officially formed in 1963 through a merger of the American Rocket Society (ARS) and the Institute of Aerospace Sciences (IAS), the purpose was, and still is, 'to advance the arts, sciences, and technology of aeronautics and astronautics, and to promote the professionalism of those engaged in these pursuits.' Both ARS and IAS brought to the relationship a long and eventful history—stretching back to 1930 and 1932, respectively—and each left its mark on the Institute. The merger combined the imaginative, opportunistic, and risk-taking desire of those rocket, missile, and space professionals with the more established, well-recognized achievers from the aviation community."

> "Today, with more than 31,000 members, AIAA is the world's largest professional society devoted to the progress of engineering and science in aviation, space, and defense. The Institute continues to be the principal voice, information resource, and publisher for aerospace engineers, scientists, managers, policymakers, students, and educators. Also, many prominent corporations and governments worldwide rely on AIAA as a stimulator of professional accomplishment in all areas related to aerospace. Consider this: Since 1963, AIAA members have achieved virtually every milestone in modern American flight."

***Key Publications:***  *Aerospace America, AIAA Bulletin, Aerospace Database, Student Journal*

### 1.2.2  Biomedical Engineering

From the Biomedical Engineering Society (*http://mecca.org/BME/BMES/society/index. htm*):

> "The Biomedical Engineering Society (BMES) is an interdisciplinary society established on February 1, 1968 in response to a manifest need to provide a society that gave equal status to representatives of both biomedical and engineering interests. As stated in the Articles of Incorporation, the purpose of the Society is: 'To promote the increase of biomedical engineering knowledge and its utilization.' Today, the society represents over 1,000 professionals and over 1,000 student members (undergraduate and graduate). There are 34 BMES student chapters and about two-thirds of the ABET-accredited bioengineering/ biomedical engineering undergraduate programs have BMES student chapters."

***Key Publications:***  *Annals of Biomedical Engineering, BMES Bulletin*

### 1.2.3 Chemical Engineering

From the American Institute of Chemical Engineers (*www.aiche.org*)

"Founded in 1908, the American Institute of Chemical Engineers (AIChE) is a nonprofit organization providing leadership to the chemical engineering profession. Representing 57,000 members in industry, academia, and government, AIChE provides forums to advance the theory and practice of the profession, upholds high professional standards and ethics, and supports excellence in education. Institute members range from undergraduate students, to entry-level engineers to chief executive officers of major corporations. As AIChE approaches the 21st Century, technological, political, social, and economic changes in our society require that we evaluate and refine our strategic plan. This will assure relevance to our members, the profession, and society at large."

"Rapid changes in skill needs and career paths of chemical engineers create new opportunities for AIChE to assist its members. Institutional stakeholders are increasingly faced with the need for effective collaborations. At the same time, the explosive change in information and communication technology introduces challenges to deliver products and services more efficiently."

"In response, our leadership has revised our vision and mission and has developed objectives and strategies that address these changes and that consider the organizational and financial resources and the business processes needed to assure AIChE's relevance in the 21st Century."

***Key Publications:*** *AIChE Journal, Chemical Engineering Progress, Environmental Progress and Process Safety Progress, Biotechnology Progress*

### 1.2.4 Civil Engineering

From the American Society of Civil Engineers (*www.asce.org*):

"Founded in 1852, the American Society of Civil Engineers (ASCE) represents more than 123,000 members of the civil engineering profession worldwide and is America's oldest national engineering society. ASCE's vision is to position engineers as global leaders building a better quality of life."

"ASCE's mission is to provide essential value to our members, their careers, our partners and the public by developing leadership, advancing technology, advocating lifelong learning and promoting the profession. From the building of the Parthenon in 432 B.C. to the building of the Petronas Towers today, the civil engineering profession has proven its sustainability. Withstanding the passage of time, civil engineers have built cultural landmarks that stand in tribute to the profession's creative spirit and ingenuity."

"Civil engineers are trained to plan, build and improve the water, sewer and transportation systems that you depend on every day. They build dams able to withstand the crushing pressure of a lake full of water. They build bridges able to resist the forces of wind and traffic. They develop environmentally friendly materials and methods, and they build things to last. So skilled is their work that we rarely stop to wonder how they design the mammoth skyscrapers we work in, the tunnels we drive in, and the stadium domes we sit beneath."

***Key Publications:*** *ASCE News, Civil Engineering Magazine*, plus numerous journals on specialized topics in civil engineering

## 1.2.5 Computer Engineering

From the Association for Computing Machinery (*www.acm.org*):

> "Founded in 1947 Association for Computing Machinery (ACM) is the world's first educational and scientific computing society. Today, our members — over 80,000 computing professionals and students worldwide — and the public turn to ACM for authoritative publications, pioneering conferences, and visionary leadership for the new millennium."

> "ACM publishes, distributes, and archives original research and first-hand perspectives from the world's leading thinkers in computing and information technologies. ACM offers over two dozen publications that help computing professionals negotiate the strategic challenges and operating problems of the day. The ACM Press Books program covers a broad spectrum of interests in computer science and engineering."

> "*Communications of the ACM* keeps information technology professionals up to date with articles spanning the full spectrum of information technologies in all fields of interest including object-oriented technology, multimedia, the Internet, and networking. *Communications* also carries case studies, practitioner-oriented articles, and regular columns, the ACM Forum, and technical correspondence."

***Key Publications:*** *The ACM Digital Library* (a collection online publications); *Communications of the ACM*; *Crossroads* (a student magazine); and various ACM Transactions journals, including: *Computer-Human Interaction, Computer Systems, Database Systems, Design Automation for Electronic System, Graphics, Information System, Mathematical Software, Modeling and Computer Simulation, Networking, Programming Languages and Systems*, and *Software Engineering and Methodology*

## 1.2.6 Electrical Engineering

From The Institute of Electrical and Electronics Engineers (*www.ieee.org*):

> "The Institute of Electrical and Electronics Engineers (IEEE) helps advance global prosperity by promoting the engineering process of creating, developing, integrating, sharing, and applying knowledge about electrical and information technologies and sciences for the benefit of humanity and the profession. IEEE provides the latest information and the best technical resources to members worldwide. Today, IEEE connects more than 350,000 professionals and students to the solutions to tomorrow's technology needs."

> "The IEEE is one of the world's largest technical professional societies. Founded in 1884 by a handful of practitioners of the new electrical engineering discipline, today's Institute is comprised of more than 350,000 members who conduct and participate in its activities in approximately 150 countries. The men and women of the IEEE are the technical and scientific professionals making the revolutionary engineering advances which are reshaping our world today."

> "The technical objectives of the IEEE focus on advancing the theory and practice of electrical, electronics, and computer engineering and computer science. To realize these objectives, the IEEE sponsors technical conferences,

symposia, and local meetings worldwide. It publishes nearly 25% of the world's technical papers in electrical, electronics, and computer engineering, and provides educational programs to keep its members' knowledge and expertise state-of-the-art."

**Key Publications:** *IEEE Spectrum; Proceedings of the IEEE*; plus over 40 specialized IEEE Transactions from its various societies, including (in alphabetical order): *Aerospace and Electronic Systems; Advanced Packaging; Antennas and Propagation; Applied Superconductivity; Automatic Control; Biomedical Engineering; Circuits and Devices; Communications; Computing in Science and Engineering Magazine (CiSE) Control Systems; Dielectrics and Electrical Insulation; Electromagnetic Compatibility; Electron Devices; Energy Conversion; Engineering Management; Geoscience and Remote Sensing; Image Processing; Industry Applications; Instrumentation and Measurement; Intelligent Systems; Lasers and Electro-Optics; Magnetics; Mechatronics; Medical Imaging; Microelectromechanical Systems; Microwave Theory and Techniques; Neural Networks; Parallel and Distributed Systems; Personal Communications; Photonics; Plasma Science; Power Electronics; Power Systems; Quantum Electronics; Reliability; Robotics and Automation; Semiconductor Manufacturing; Signal Processing; Software Engineering; Solid-State Circuits; Systems, Man, and Cybernetics; Ultrasonics, Ferroelectrics, and Frequency Control; Vehicular Technology; Very Large Scale Integration Systems*; and *Visualization and Computer Graphics*

### 1.2.7 Industrial and Manufacturing Engineering

From the Institute of Industrial Engineers (*www.iienet.org*):

> "Founded in 1948, the Institute of Industrial Engineers is the society dedicated to serving the professional needs of industrial engineers and all individuals involved with improving quality and productivity. Its 24,000 members throughout North America and more than 80 countries stay on the cutting edge of their profession through IIE's life-long-learning approach, as reflected in the organization's educational opportunities, publications, and networking opportunities. Members also gain valuable leadership experience and enjoy peer recognition through numerous volunteer opportunities."

**Key Publications:** *IIE Solutions, Industrial Management, IIE Transactions, The Engineering Economist, Student IE, Journal of the Society for Health Systems*

### 1.2.8 Mechanical Engineering

From the American Society of Mechanical Engineers (*www.asme.org*):

> "The 125,000-member ASME International is a worldwide engineering society. It conducts one of the world's largest technical publishing operations, holds some 30 technical conferences and 200 professional development courses each year, and sets many industrial and manufacturing standards. Founded in 1880 as the American Society of Mechanical Engineers, today ASME International is a nonprofit educational and technical organization serving a worldwide membership."

> "The work of the Society is performed by its member-elected Board of Governors and through its five Councils, 44 Boards, and hundreds of Committees

in 13 regions throughout the world. There are a combined 400 sections and student sections serving ASME's worldwide membership."

"Its vision is to be the premier organization for promoting the art, science and practice of mechanical engineering throughout the world. Its mission is to promote and enhance the technical competency and professional well-being of our members, and through quality programs and activities in mechanical engineering, better enable its practitioners to contribute to the well-being of humankind."

*Key Publications: Mechanical Engineering; ASME News; Applied Mechanics Reviews; plus journals in numerous specialty areas; including Applied Mechanics; Heat Transfer; Biomechanical Engineering; Computing and Information Science; Dynamic Systems; Measurement and Control; Electronic Packaging Energy Turbines and Power Engineering Materials Technology Fluids; Manufacturing Science; Mechanical Design; Offshore Mechanics; Arctic Engineering; Pressure Vessel Technology; Solar Energy; Tribology; Turbomachinery; Vibration and Acoustics; and Mechatronics*

## PROFESSIONAL SUCCESS: CHOOSING A FIELD OF ENGINEERING

If you are a first-year student of engineering, you may already have decided upon a major field. After taking several required courses, however, you may not be sure if you've chosen the right type of engineering. Conversely, you may have entered school without committing yourself to any one field of engineering. If you find yourself in either of these situations, you're probably wondering how one goes about choosing a career direction in engineering.

One way to find out more about the different branches of engineering is to attend technical talks and seminars hosted by the engineering departments in your college or university. Such talks are usually aimed at graduate students and faculty, so much of the material will be over your head. Simply *exposing* yourself to these technical talks, however, will give you a feeling for the various branches of engineering and help you find one that most closely matches your skills and interests.

Most schools host workshops in career advising. Be sure to attend one. Talk with the experts in career planning and job placement. Many college campuses host student chapters of professional organizations. These groups often organize tours of engineering companies. Attending such a tour can provide valuable perspective about the activities of a particular branch of engineering and provide you with an idea about what life as an engineer will be like.

One of the most valuable resources for career advice is your own college faculty. Get advice from your advisor about which major is right for you. Also, professors love to talk about their work. Invite a professor to your dormitory or living unit to speak to students about choosing an engineering career. Speak to your department about hosting a career night in which a panel of professors answers questions about jobs in engineering. Learn to make use of all available resources for help in choosing your college major.

## 1.3   THE ENGINEER: CENTRAL TO PROJECT MANAGEMENT

When we think of the word "design," we may imagine a lone engineer sitting in a cubicle at a computer terminal, or perhaps in a workshop, crafting some marvelous piece of technical wizardry. As a student, you may be eager to pursue this notion of the rugged individual—the sole entrepreneur who single-handedly changes the face of technology. You might ask, "Why do I have to take all of these *other* courses? Why can't I just take courses that are of interest to me or important to my career goals?" The answer to these questions lies in the multidisciplinary nature of engineering. At times, an engineer does work alone, but most of the time, engineers must interface with individuals who come from different educational backgrounds. Engineering projects can be complex undertakings that require teamwork and the coordination of many people of different skills and personality traits. An engineer must learn the languages of physicists, mathematicians, chemists, managers, fabricators, technicians, lawyers, marketing staff, and secretaries. It's been said that a good engineer acts as the glue that ties a project together, because he or she has learned to communicate with specialists from each of these varied fields. Learning to communicate across all these occupations requires that the engineer have a broad education and the ability to apply a full range of skills and knowledge to the design process.

### 1.3.1 The Well-Rounded Engineer

To help illustrate the breadth of communication skills required of an engineer, imagine that you work for the fictitious company depicted in Figure 1.8. Each person shown in the outer circle brings to the company a different professional expertise and is represented by a famous person with an appropriate background. Notice that you, the design engineer, are in the center of the organizational circle. Other engineers on your design team may join you in the center, but each of you can easily communicate with any one specialist in the outer ring. As an engineer, you've taken courses or have been exposed to each of their various disciplines. This unique feature of your educational background enables you to communicate with anyone in the professional circle and positions you as the individual most likely to act as central coordinator.

***The Physicist*** (*e.g., Albert Einstein, best known for his theory of relativity*).    The physicist of the company is responsible for understanding the basic physical principles that underlie the company's product line. He spends his time in the laboratory exploring new materials, analyzing their interactions with heat, light, and electromagnetic radiation. He may discover a previously unknown quantum interaction that will lead to a new semiconductor device or perhaps he will explore the potential for using superconductors in the company's product. Or, he may simply perform the physical analysis for a new micro-accelerometer. Because you've taken two or more semesters of basic physics and have learned some mechanics, thermodynamics, and electromagnetics, you can easily converse with the physicist and discuss how his basic discoveries relate to the practical interests of the company.

***The Chemist*** (*e.g., Marie Curie, who discovered radium*).    The chemist analyzes materials and substances used in producing company products. She ensures that raw materials used for manufacturing meet purity specifications so that quality control can be maintained. In her laboratory, she directs a team of experimentalists who seek to discover improved materials that are stronger and more durable than those currently being used. She may perform research on complex organic compounds or perhaps work on molecular-based nanotechnology. As an engineer, you've taken one or more courses in chemistry and can speak her language. You understand such concepts as reaction rates, chemical equilibrium, molarity, reduction and oxidation, acids and bases, and electrochemical potential. Perhaps you're a software engineer writing a program that will control a chemical analysis instrument. Maybe you are a manufacturing engineer charged with translating a chemical reaction into a manufactured product. Whatever your role, you are an individual very well suited to bringing the contributions of the chemist to the design process.

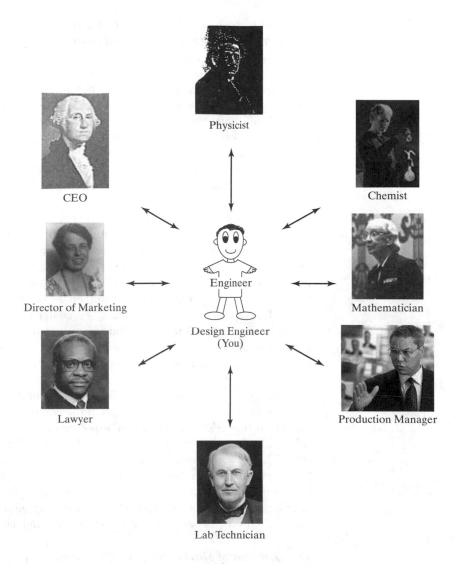

***Figure 1.8.*** The professional circle with the design engineer at its center.

***The Mathematician** (e.g., Grace Hopper, former Navy admiral, mathematician, and computer specialist responsible for the term "computer bug").* The mathematician of the company, who might also be a computer scientist, worries about things such as modeling, statistics, databases, and forecasting. She may be involved in an intriguing new database algorithm or mathematical method for modeling an engineering system. Perhaps she uses mathematics to analyze the company's production line or to forecast trends in marketing. You converse easily with the mathematician, because you have taken numerous math courses as part of your engineering program. Although your emphasis has been on applied, rather than pure mathematics, you're familiar with calculus, differential equations, linear algebra, statistics, probability, vector algebra, and complex variables. You can easily apply the concepts of mathematics to problems in engineering design.

***The Production Manager*** (*e.g., Colin Powell, U.S. Secretary of State, former U.S. Army general, military planner, and co-architect of Operation Desert Storm*).    Like the army general in top command, the production manager is responsible for mobilizing materials, supplies, and personnel to manufacture company products. The production manager may worry about things such as job scheduling, quality control, materials allocation, quality assurance testing, and yield. As the engineer who designs products, you work closely with the production manager to make sure that your design approach is compatible with the company's manufacturing capabilities. Your training as an engineer and your exposure to machining, welding, circuit fabrication, and automation has given you the ability to understand the job of the production manager and has provided you with the vocabulary needed to communicate with him.

***The Lab Technician*** (*e.g., Thomas Edison, famous tinkerer and experimenter, best known for inventing the incandescent light bulb*).    The lab technician is an indispensable member of the design team. An habitual tinkerer and experimenter, the lab technician helps bring your design product to fruition. He is adept at using tools and has much knowledge about the practical aspects of engineering. The lab technician is masterful at fabricating prototypes and is likely to be the individual who sets up and tests them. The typical lab technician has a degree in engineering technology, hence you and he have taken many of the same courses, although your courses probably have included more formal theory and mathematics than his. You communicate easily with the lab technician and include him in each phase of your design project.

***The Lawyer*** (*e.g., Clarence Thomas, lawyer and Supreme Court Justice*).    The lawyer worries about the legal aspects of the company's products. Should we apply for a patent on the XYZ widget? Are we exposing ourselves to a liability suit if we market a substandard product? Is our new deal with Apex Corporation fair to both companies from a legal perspective? To help the lawyer answer these questions, you must be able to communicate with him and share your engineering knowledge. The logical thought that forms the basis of law is similar to the methods you've used to solve countless engineering problems. As an engineer, you easily engage in discourse with the lawyer and can apply his legal concerns about safety, ethics, and liability to the design process.

***The Director of Marketing*** (*e.g., Eleanor Roosevelt, former First Lady of the United States*).    The director of marketing is a master of imagery and style. Her job is to sell the company's products to the public and convince people that your products are better than those of your competitors. The marketing manager has excellent communication skills, some knowledge of economics, and an understanding of what makes people want to buy. You interface easily with the marketing manager because you've dealt with all aspects of design as part of your training as an engineer. Through this training, you have focused not only on technical issues, but also on things such as product appearance, the human-machine interface, durability, safety, and ease of use. Your familiarity with these important issues has prepared you to help the director of marketing understand your product and how it works. You can respond to her concerns about what the public needs from the product that you design.

***The President/Chief Executive Officer*** (*e.g., George Washington, first president of the United States*).    The CEO of the company probably has an MBA (Master's of Business Administration) or higher degree and a long history working in corporate financial affairs. The CEO worries about the economy and what future markets the company should pursue or whether to open a new plant in a foreign country. It's the

CEO who determines how your current project will be financed, and he needs to be kept up to date about its progress. The CEO also may ask you to assess the feasibility of a new technology or product concept. As an engineer, you have no difficulty conversing with the CEO, because the economic principles of profit and loss, cost derivatives, statistics, and forecasting are closely tied to concepts you learned in courses on calculus, statistics, and economics. You've learned to use spreadsheets in one or more engineering classes and have no trouble interpreting or providing the information that is part of the CEO's world. Likewise, your training as an engineer prepares you to communicate with the CEO about the impact of your design project on the economic health of the company.

## 1.4  ENGINEERING: A SET OF SKILLS

To be successful at design, an engineer must acquire technical, theoretical, and practical competency and must be a good organizer, and communicator. Three especially important skills that are at the foundation of engineering design are *knowledge*, *experience*, and *intuition*. These talents do not form an exhaustive set, but they are crucial to the well-rounded engineer.

### 1.4.1   Knowledge

*Knowledge* describes the body of facts, scientific principles, and mathematical tools that an engineer uses to form strategies, analyze systems, or predict results. An engineer's acquired knowledge can provide a deeper understanding of how something works. The natural sciences, for example physics, chemistry, and biology, help an engineer understand the physical world. Mathematics provides a universal technical language that spans various disciplines and can be understood by anyone regardless of spoken language or cultural background. Each field of engineering has its own traditional body of knowledge, but an engineer in one field also learns subjects from other fields. Areas of knowledge that are common to all engineers include mechanics, circuits, materials science, and computer programming.

As a student of engineering, you may ask why you are required to take subjects that seem irrelevant to your career aspirations. Any experienced engineer will tell you the answer: Engineers work in a multidisciplinary world where basic knowledge of many different subjects is necessary. Mechanical and computer engineers use electrical circuits. Electrical engineers build physical structures and use computers. Aeronautical engineers rely on software systems. Software engineers design airplane controls. Understanding the field of another engineer is critical to cross-disciplinary communication and design proficiency.

Although formal education is an important part of any engineer's training, the prudent engineer also acquires knowledge through on-the-job training and a lifetime of study and exploration. Tinkering, fixing, experimenting, and taking things apart to see how they work also are important sources of engineering knowledge. As a young person, did you disassemble your toys, put together model kits, write your own computer games, create Web pages, or play with building sets, hammers, nails, radios, bicycles, or computers? Without knowing it, you began the path toward acquiring engineering knowledge. The professional engineer engages in similar practices. By becoming involved in all aspects of a design project, by keeping up to date with the latest technology, by taking professional development courses and solving real-world problems, the practicing engineer remains current and competent.

### 1.4.2   Experience

*Experience* refers to the body of methods, procedures, techniques, and rules of thumb that an engineer uses to solve problems. For the engineer, accumulating experience is just as important as acquiring knowledge. As a student, you will have several opportunities to gain engineering experience. Cooperative assignments, assistantships in labs, capstone design projects, summer jobs, and research work in a professor's laboratory provide important sources of engineering experience. On-the-job training is also a good way to gain valuable professional experience. Many engineering companies recognize this need and provide entry-level engineers with initial training as a way of infusing additional experience. Developing experience requires "seasoning," the process by which a novice engineer gradually learns the "tricks of the trade" from other, more experienced engineers. Company lore about methods, procedures, and history is often passed orally, from one engineer to the another, and a new engineer learns this information by working with other engineers. The history of what *hasn't* worked in the past is also a key part of this oral tradition.

An engineer also gains valuable experience by enduring design *failure*. When the first attempt at a design fails in the testing phase, the wise engineer views it as a learning experience and uses the information to make needed changes and alterations. Experience is acquired by testing prototypes, studying failures, and observing the results of design decisions.

Engineers also must consider the issues of reliability, cost, manufacturability, ergonomics, and marketability when making design decisions. Only by confronting these constraints in real world situations can an engineer gain true design experience.

### 1.4.3   Intuition

*Intuition* is a characteristic normally associated with fishermen, fortunetellers, and weather forecasters. Intuition is also an essential element of engineering. It refers to an engineer's basic instinct about what will or will not work as a problem solution. Although intuition can never replace careful planning, analysis, and testing, it can help an engineer decide which approach to follow when faced with many choices and no obvious answer. An intuitive feeling for what will work and what will not work, grounded in extensive experience, can save time by helping an engineer choose the path that will eventually lead to success rather than failure. When intuition is at work, you may hear an engineer using phrases such as, "That seems reasonable" or "That looks about right." or "Oh, about this much."

Intuition is a direct by-product of design experience and is acquired only through practice, practice, and more practice. In the information age, where much of engineering focuses on computers, engineers are tempted to solve everything by simulation and computer modeling. While the use of computers has dramatically accelerated the design cycle and has dramatically changed the practice of engineering, using them makes it easy to forget that a product ultimately must obey the idiosyncrasies of the real physical world. Developing intuition about that world is an important part of your engineering education. The difference between a good engineer and an excellent one is often just an instinct for how the laws of nature will manifest themselves in the design process. Will too much heat overpower that circuit? Will friction rob that engine of too much power? How big should the vessel be for a production run of that new cosmetic? Developing intuition should be a key goal of your engineering education. How many times have you opened your computer to install new components? Do you alter hardware settings just to see what happens? Have you opened the hood of family car just to see what lies beneath it? Have you adjusted the gears on your bicycle? Have you put together a kit or built your science project from raw materials? Each of these tasks helps you acquire intuition. Observing the way in which other engineers have laid out the boards of a computer will acquaint you with the techniques of hardware design. Adjusting the gear and brake setting of your bicycle will help you to understand design trade-offs, such as the conflict between strength and durability versus lightweight construction. Becoming knowledgeable in the use of tools will help you to better understand the impact of your design decisions on manufacturing. Repetition, testing, careful attention to detail, working with more experienced engineers, and dedication to your discipline are the keys to developing design intuition. Design intuition is best acquired by "doing design," that is, by playing with real things.

**PROFESSIONAL SUCCESS: HOW TO GAIN EXPERIENCE AS A STUDENT**

Your experience as an engineer can begin while you are a student. If your school has one, a cooperative education program is an excellent way to gain experience as an engineer. The typical program places you as an intern in an engineering company for 6 to 12 months. You'll typically be assigned to a senior engineer to assist in such tasks as computer-aided design, software development, product prototyping, testing, laboratory evaluation, or other work. You'll get to see how the company works, and the company will get to evaluate you as a possible future hire. In addition, you'll be paid for the time you spend at the company.

Students also can gain valuable experience by working in research labs at school. Most professors are delighted to take eager undergraduates into their research laboratories. Most schools list the research interests of the faculty on departmental Web pages. Learn about the research activities of a professor whose class you have enjoyed. Don't be afraid to simply ask if she needs help in the lab. Many professors receive industry or government funding for their research, and many sponsor "Research for Undergraduates" programs, so you may even be paid an hourly wage or small stipend for your time. You'll be assigned tasks such as constructing experiments, wiring circuits, writing programs, obtaining data, preparing test samples, or assisting graduate students.

## KEY TERMS

Engineering
Experience
Career

Profession
Intuition

Knowledge
Management

# 2

# What Is Design?

**design n:**  Invention and disposition of the form, parts, or details of something according to a plan.[1]

Engineers have made tremendous contributions to the quality of life in the 20th and 21st centuries. The automobile, bicycle, airplane, transatlantic cable, heart pacemaker, radio and television, international shipping, national highway system, national power grid, personal computer, Internet, cellular telephone, fax machine, CD player, and global positioning system all began with the common thread of *design*. From a fundamental perspective, design can be defined as any activity that results in the synthesis of something that meets a need. A refrigerator keeps food cold; a bicycle provides transportation; a keyboard sends data to a computer; a muffler silences a noisy automobile. Although design is practiced by all sorts of individuals, technical and nontechnical alike, the notion of "design" in the context of engineering implies the application of knowledge and specialized skills toward the creation of something that meets a desired set of specifications. The object of design might be a device, machine, circuit, building, mechanism, structure, software program, operating system, manufacturing process, or other technical system. In an engineering context, the word "design" answers the simple question, "What do engineers do?"

## SECTIONS

- 2.1    The Use of the Word "Design"
- 2.2    The Difference Between Design, Analysis, and Reproduction
- 2.3    Good Design Versus Bad Design
- 2.4    The Design Cycle
- 2.5    A Design Example

## OBJECTIVES

*In this chapter, you will learn about:*

- The meaning of the word "design."
- The difference between design, analysis, and reproduction.
- The difference between good and bad design.
- The engineering design process.
- The design cycle.
- A design competition example.

---

[1] *Webster's II New College Dictionary*. Boston: Houghton-Mifflin Company, 1995, p. 307.

## 2.1   THE USE OF THE WORD "DESIGN"

In this book, the word "design" will be used in several ways. It may be used as a verb, as in, "Design a widget that can open a soda can automatically." The word might be used as a noun that defines the creation process itself, as in, "Learning design is an important part of engineering education." Alternatively, design may be used as a noun that describes the end result of the process, as in "The design was a success and met the customer's specifications." The word may be used as an adjective, as in, "This book will help you learn the design process."

Sometimes, an alternative word is needed to describe the end result of a design effort. For this purpose, the word "product" may be used in its generic sense, even if the thing being designed is not a product for sale. Similarly, the word "device" may be used to describe the results of a design effort, even if the entity is not a physical apparatus. Thus, the words product and device can refer not only to tangible objects, but also to large structures, systems, procedures, or software.

## 2.2   THE DIFFERENCE BETWEEN DESIGN, ANALYSIS, AND REPRODUCTION

Students of engineering often are confused by the distinction between analysis, reproduction, and design. In science classes, students are asked to answer problems, observe phenomena in the lab, and perform calculations. In engineering classes, instructors instead may stress the importance of design. The difference between analysis and design can be defined in the following way: If only one answer to the problem exists, and finding it merely involves putting together the pieces of the puzzle, then the activity is probably analysis. For example, processing data and using it to test a theory is analysis. On the other hand, if more than one solution exists, and if deciding upon a suitable path demands being creative, making choices, performing tests, iterating, and evaluating, then the activity is most certainly design. Design can include analysis, but it also must involve at least one of these latter elements.

As an example of the distinction between analysis and design, consider the weather station shown in Figure 2.1. This remote-controlled buoy is located off the coast of California and is maintained by the U.S. National Oceanic and Atmospheric Administration (NOAA). It provides 24-hour data to mariners, the Coast Guard, and weather forecasters. Processing the data stream from this buoy, posting it on the Internet, and using the information to forecast the weather are examples of analysis. Deciding *how* to build the buoy so that it meets the needs of NOAA is an example of design.

Another example that illustrates the difference between analysis and design can be found in the medieval catapult of Figure 2.2. Determining the projectile's $x$-$y$ trajectory is an analysis problem that involves computing the $x$ and $y$ components of Newton's law of motion $\mathbf{a} = \mathbf{F}/m$ that is,

$$\frac{d^2x}{dt^2} = \frac{dv_x}{dt} = 0 \qquad (2\text{-}1)$$

and

**Figure 2.1.** Automated weather buoy located off the coast of California provides information about local sea and weather conditions. (*Photo courtesy of National Data Buoy Center.*)

$$\frac{d^2y}{dt^2} = \frac{dv_y}{dt} = \frac{-mg}{m} = -g, \qquad (2\text{-}2)$$

where $v_x$ and $v_y$ are the $x$- and $y$-components of the projectile's velocity, respectively. Here, $g$ is the gravitational constant (9.8 m/s$^2$ in MKS units); $m$ the mass of the projectile (e.g., a stone); and $mg$ the gravitational force on the projectile in the $-y$-direction. The second derivatives $d^2x/dt^2$ and $d^2y/dt^2$ represent the $x$- and $y$-components of the projectile's acceleration. Note that the projectile experiences no $x$-directed force after its initial launch; hence, $v_x$ will remain unchanged for $t > 0$.

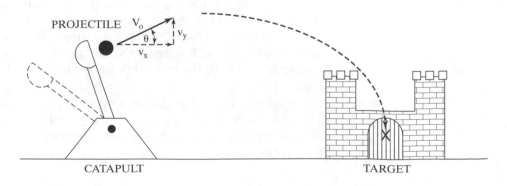

**Figure 2.2.** A catapult sends a projectile along a trajectory toward a target. Computing the parameters needed to hit the target is an example of analysis. Determining how to best build the catapult is an example of design.

Equations (2.1) and (2.2) can be solved only if the initial values of $V_o$ and $\theta$, the so-called *initial conditions* at $t = 0$, are specified. The initial values of $V_o$ and $\theta$ can be used to find initial conditions for the $x$- and $y$-components of velocity as well:

$$\frac{dx}{dt} = v_x = V_0 \cos\theta \tag{2-3}$$

and

$$\frac{dy}{dt} = v_y = V_0 \sin\theta, \tag{2-4}$$

where $V_o$ is related to $v_x$ and $v_y$ by

$$V_0 = [\, v_x^2 + v_y^2 \,]^{1/2}. \tag{2-5}$$

The launch speed $V_0$ and launch angle $\theta$ of the projectile are set by the user of the cata-pult. As suggested by Figure 2.2, the user must first choose the target point, then adjust $V_0$ and $\theta$ so that the stone hits the desired target. Although this process involves making decisions (*Which target shall I hit?*) and setting parameters (*Which $V_0$ and $\theta$ shall I choose?*), and although the problem has more than one possible solution, it requires analysis only and involves no design.

In contrast, determining *how to build* the catapult most certainly involves design. The machine can be built in more than one way, and the designer must decide which method is best. Should the structural members be made from oak or pine? Should twisted ropes, a hanging basket of rocks, or bent branches be used as the energy storage mechanism? How large should the machine be? Should it rest on wheels or skids? Answering these questions requires experimentation, analysis, testing, evaluation, and revision, which are all elements of the design process.

This example illustrates the difference between design and analysis. Design involves the creation of a device or product to meet a need or set of specifications. Analy-sis refers to the process of applying mathematics and other tools to find the answer to a problem. In contrast to both, the word *reproduction* refers to the process of recreating something that has already been designed. Reproduction may involve an exact replication, or it may involve minor revisions whose consequences have already been determined. For instance, copying an oscillator circuit from an electronics book and substituting resistor values to set the frequency is an example of reproduction rather than design. Similarly, building a backyard shed from a set of purchased plans involves reproduction, but no design. Reproduction is an important part of engineering and lies at the core of manufac-turing, but it does not require the same set of skills and tools as does true design.

**EXAMPLE 2.1: TRAJECTORY ANALYSIS**

Equations (2.1) through (2.4) form a set of *differential equations* that describe the motion of the catapult's projectile. Given an initial $V_0$ and $\theta$, these equations can be solved for the projectile's position coordinates $x(t)$ and $y(t)$ as functions of time. The techniques for obtaining such a solution are covered in math courses taken by most stu-dents of engineering; however, many first-year students do not yet have the math skills needed to solve differential equations. Another method by which solutions for the pro-jectile coordinates $x(t)$ and $y(t)$ can be found as functions of time is to find them iteratively by computer. The flowchart of Figure 2.3 illustrates the basic roadmap for writing such a program.

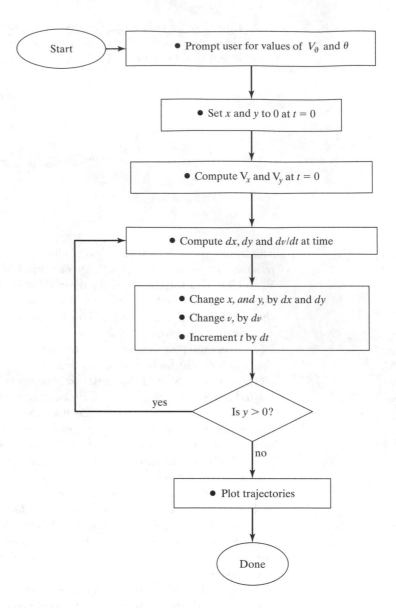

**Figure 2.3.** Flowchart for the iterative solutions of Eqs. (2.1) through (2.4), given the initial conditions $V_o$ and $\theta$.

The listing shown below implements the program described by the flowchart of Figure 2.3 using the programming language MATLAB™. Many other programming languages can be used as well, but MATLAB is chosen here because of its widespread use in the engineering community and the generic nature of its commands and functions. MATLAB code is case sensitive. The percent symbols (%) denote comment lines, and the asterisks (∗) denote multiplication. A semicolon (;) at the end of a line simply tells the program not to display the result of that particular calculation on the screen. The plot of $x$ versus $y$ for the initial conditions $V_o = 15$ m/s and $\theta = 60°$ is shown in Figure 2.4.

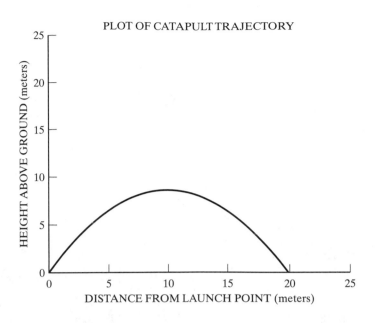

**Figure 2.4.** Result of the computation of Figure 2.3 for the case $V_o$ = 15 m/s and $\theta$ = 60°.

```
%%% MATLAB PROGRAM CODE %%%
%THIS PROGRAM CALCULATES and PLOTS the TRAJECTORY of a
CATAPULT PROJECTILE

%Prompt user for initial conditions:
Vo = input('Enter value of projectile launch velocity in
m/s ')
theta = input('Enter value of launch angle in degrees from
horizontal ')

%Convert launch angle from degrees to radians:
theta = (theta/180)*pi;

%set x and y to zero at t=0:
x=0; y=0; t=0;
%set value of time increment dt to 10 milliseconds:
dt=1e-2;
%set gravitational constant in meters/sec^2 (MKS units):
g=9.8;

%Compute x- and y-components of launch velocity at t=0:
vx = Vo*cos(theta);
vy = Vo*sin(theta);

%Declare initially empty vectors X, Y, and T
%(A construct in MATLAB for storing calculated plots for
latter plotting)
X=[ ]; Y=[ ]; T=[ ];

%THE LOOP BEGINS HERE: Iterate around the flow-chart loop
%while y is greater than zero
while y >= 0
```

```
%compute incremental changes to x, y, and vy at time t
dx = vx*dt;
dy = vy*dt;
dvy = -g*dt;%downward acceleration due to gravity

%update values of x, y,vy, and t:
x = x + dx;
y = y + dy;
vy = vy + dvy;
t = t + dt;

%save updated values for plotting when iteration is
finished
X = [X x];
Y = [Y y];
T = [T t];

end  %THE LOOP ITERATES HERE (exit if y is not greater than
zero)

%When calculations are finished, plot the results:
close;
axis([0 100 0 100]); axis manual; hold on;
%Set graph axes to fixed values
plot(X,Y);%plot trajectory of projectile (x and y values)
title('PLOT OF CATAPULT TRAJECTORY')
xlabel('DISTANCE FROM LAUNCH POINT (meters)')
ylabel('HEIGHT ABOVE GROUND (meters)')
```

## PRACTICE!

For Exercises 1 to 7, determine whether the following tasks involve analysis, design, or reproduction:

1. Find the best travel route between Chicago and Houston.

2. Find a way to prevent people from burning their hands while removing a bagel from a toaster oven.

3. Find the best dimensions of a 16-oz soup can so that a packing box for 24 cans has the smallest volume.

4. Find a way to mount a cellular telephone on a bicycle to permit safe, hands-free operation.

5. Find a way to produce bar-coded badges that can be handed out to the attendees of a technical conference.

6. Find a way to produce 1,200 origami (folded paper) nut containers for a large alumni dinner.

7. Find a way to use a global positioning system (GPS) receiver to automatically navigate a lawn mower around a nonrectangular yard.

8. *Challenge Exercise*: Show by direct solution that Equations (2.1) through (2.4) yield a parabolic trajectory.

9.  *Challenge Exercise*: A catapult target lies 100 m away. Determine at least one possible set of values for $V_0$ and $\theta$ in Figure 2.2 so that a 5-kg stone hits the target. How much energy must the catapult impart to the stone to meet this objective? (Hint: the kinetic energy of the stone when it first leaves the catapult is $mV_0^2/2$.)

10. Modify the MATLAB program shown above so that it plots (a) the height of the projectile as a function of time, and (b) the angle of the trajectory relative to the horizontal as a function of time.

## 2.3 GOOD DESIGN VERSUS BAD DESIGN

Anyone who has taken a car in for repair recognizes the difference between a good mechanic and a bad mechanic. A good mechanic diagnoses your problem in a timely manner, fixes what's broken at a fair price, and makes repairs that last. A bad mechanic fails to find the real problem, masks the symptoms with expensive solutions that don't last, and charges too much money for needless repairs. Engineers are a bit like auto mechanics in this respect. The world is full of both good engineers and bad engineers. Just because an engineer has produced something does not mean that the product has been designed well. Just because the design works initially doesn't mean that the product will last over time. Although the criteria by which a product is judged varies with the nature of the product, the success of most design efforts can be judged by the general characteristics summarized in Table 2-1

**TABLE 2-1**  Characteristics of Good Design Versus Bad Design

| GOOD DESIGN | BAD DESIGN |
|---|---|
| 1. Meets all technical requirements | 1. Meets only some technical requirements |
| 2. Works all the time | 2. Works initially but stops working after a short time |
| 3. Meets cost requirements | 3. Costs more than it should |
| 4. Requires little or no maintenance | 4. Requires frequent maintenance |
| 5. Is safe | 5. Poses a hazard to users |
| 6. Creates no ethical dilemma | 6. Raises ethical questions |

The contrast between good and bad design is readily illustrated by the catapult of Figure 2.5. Suppose that the Apex Catapult Corporation has been asked to produce this device (actually called a *trebuchet*) for a brigade intent on recapturing their castle. The buyers will judge the worthiness of the catapult based on the considerations outlined in Table 2-1, as illustrated by the following discussion.

### 1. Does the Product Meet Technical Requirements?

It might seem a simple matter to decide whether or not a catapult meets its technical requirements. Either the stone hits its target, or it does not. But success can be judged in many ways. A well-designed catapult will accommodate a wide range of stone weights, textures, and sizes. It will require the efforts of only one or two people to operate, and will repeatedly hit its target, even in strong wind or rain. A poorly designed catapult may meet its launch specification under ideal conditions, but it may accommodate stones of only a single weight or require that only smooth, hard-to-find stones be used. It may not work in the rain, or it may not produce repeatable trajectories. When the arm of a poorly designed catapult is released, it may hit its own support structure, causing the stone to lose momentum and fall short of the target. The catapult might work fine for the first few launches, only to fail at a later time.

## 2. Does the Product Work?

During the development stage, a product need not be "bug"-free the very first time it is tested. However, it *must* work perfectly before it can be delivered to the customer. It must be durable and not fail after only a short time in the field. The catapult of Figure 2.5 provides an excellent example of this second principle. Even a bad designer could produce a catapult capable of meeting its specifications upon initial delivery. The Apex Corporation could make the catapult from whatever local timbers were available.

**Figure 2.5.** Reproduction of a medieval catapult called the "trebuchet." (*Photo courtesy of Middelaltercentret.*)

It might use a simple trigger mechanism made from vines and twigs. The bad designer would build the catapult as he went along, adding new features on top of old ones without examining how each feature interacted with those before it. The catapult would likely pass inspection upon delivery and be able to hurl stones several times before fraying a line, cracking a timber, or breaking its trigger mechanism. After a short period of use, however, the ill-designed timbers of its launch arm might weaken, causing the projectile to fall short of its target.

A good designer would develop a robust catapult capable of many long hours of service. This conscientious engineer would test different building materials, carriage configurations, trigger mechanisms, and launch arms before choosing materials and design strategies. The catapult would be designed as a whole, with consideration given to how its various parts interacted. The process typically would require stronger and more expensive materials, but it would prove more reliable and enable the user to hit the target repeatedly.

### 3. Does the Product Meet Cost Requirements?

Some design problems can be approached without regard to cost, but in most cases, cost is a major factor in making design decisions. Often a trade-off exists between adding features and adding cost. A catapult made from cheap local wood will be much less expensive than one requiring stronger, imported wood. Will the consumer be willing to pay Apex a higher price for a stronger catapult? Durable leather thongs will last longer than links made of less expensive hemp rope. Will the consumer absorb the cost of the more durable thongs? Painting the catapult will make it visually more attractive but will not enhance performance. Will the customer want an attractive piece of machinery at a higher price? An engineer must face questions such as these in just about every design situation.

### 4. Will the Product Require Extensive Maintenance?

A durable product will provide many years of flawless service. Durability is something that must be planned for as part of the design process, even when the cost of the final product is important. At each step, the designer must decide whether cutting corners to save money or time will lead to component failure later on. A good designer will eliminate as many latent weaknesses as possible. A bad designer will ignore them as long as the product can pass its initial inspection tests. If the Apex Catapult Corporation wishes to make a long-lasting product worthy of its company name, then it will design durability into its catapult from the very beginning of the design process.

### 5. Is the Product Safe?

Safety is a quality measured only in relative terms. No product can be made completely hazard free, so when we say that a product is "safe," we mean that it has a significantly smaller probability of causing injury than does a product that is "unsafe." Assigning a safety value to a product is one of the harder aspects of engineering design, because adding safety features usually requires adding cost. Also, accidents are subject to chance, and it can be difficult to identify a potential hazard until an accident occurs. An unsafe product may never cause harm to any one user, while statistically, some fraction of a large group of users is likely to sustain injury. The catapult provides an example of the trade-off between safety versus cost. Can a catapult be designed that provides a strategic advantage without injuring people? When a stone is thrown at the door of a castle, a probability exists that it will hit a person instead. Designing a device that can throw, say, large bags of water instead of stones would reduce the potential for human injury, but at the added cost of producing water bags. Features also could be added to the catapult to protect its users. Guards, safety shields, and interlocks would prevent accidental misfirings, but would add cost and inconvenience to the finished product.

### 6. Does the Product Create an Ethical Dilemma?

The catapult has been chosen as an example for this section because it poses a common ethical dilemma faced by engineers: Should a device be built simply because it *can* be built? A catapult, for example, can be a lethal device. When asked to build a catapult, is Apex obligated to build it? Is Apex responsible for suggesting alternatives to the rescue brigade? A less destructive battering ram might help recapture the castle while sparing innocent lives. Quiet diplomacy in lieu of force may lead to resolution and peaceful cooperation. As contrived as this fictitious example may be, it exemplifies the ethical dilemmas that may confront you as an engineer. If asked by a future employer to design offensive military weapons, will you find it personally objectionable? If your boss asks you to use cheaper materials but bill the customer for more expensive ones, will you comply with these instructions or defy your employer? If you discover a serious safety flaw in your company's product that might lead to human injury, will you insist on costly revisions that will reduce the profitability of the product? Or will you say nothing and hope for the best? Questions of these sorts are never simple to answer, but engineers face them regularly. As part of your training as an engineer, you must learn to apply your own ethical standards, whatever they may be, to problems that you encounter on the job. This aspect of design will be one of the hardest to learn, but it is one that you must master if you wish to be an engineer.

### PROFESSIONAL SUCCESS: CHOOSE A GOOD DESIGNER TO BE YOUR MENTOR

There is a difference between good designers and bad designers. Practicing engineers of both types can be found in the engineering profession, and it's up to you to learn to distinguish between the two. As you make the transition from student to professional engineer, you are likely to seek a mentor at some point in your career. Be certain that the individual you choose follows good design practices. Seek an engineer who has an intrinsic feeling for why and how things work. Find someone who adheres to ethical standards that are consistent with your own. Avoid "formula pluggers" who memorize equations and blindly plug in numbers to arrive at design decisions but have little feeling for what the formulas actually mean. Avoid engineers who lack vision and perspective. Likewise, shun engineers who take irresponsible shortcuts, ignore safety concerns, or choose design solutions without thorough testing. In contrast, do emulate engineers who are well respected, experienced, and practiced at design.

## 2.4    THE DESIGN CYCLE

Design is an iterative process. Seldom does a finished product emerge from the design process without undergoing changes along the way. Sometimes, an entire design approach must be abandoned and the product redesigned from the ground up. The sequence of events leading from idea to finished product is called the *design cycle*. Although the specific steps of the design cycle may vary with the product and field of engineering, most cycles resemble the sequence depicted in Figure 2.6. The following section explores this diagram in more detail.

### 2.4.1    Define the Overall Objectives

You should begin any new project by defining your design objectives. This step may seem a nuisance to the student eager to build, test, or write software, but it is one of the most important steps. Only by viewing the requirements from a broad perspective can an engineer determine all factors relevant to the design effort. Good design involves more than just making technical choices. The engineer must ask the following questions: Who will use the product? What are the needs of the end user? What will the product look like? Which of its features are critical, and which are only desirable? Can the product be manufactured easily? How much will it cost? What are the safety factors? Who will decide how much risk is acceptable? Answering these questions and similar ones at the outset will help at each subsequent stage of the design process.

### 2.4.2    Gather Information

In the early stages of a new project, much time should be devoted to gathering information. Learn as much as possible about related technology. Identify off-the-shelf items, systems, or software components that can be incorporated directly into your design, so that you don't have to "reinvent the wheel." Look for product descriptions, data sheets, and application notes on the Web. Keep this information in a file folder (either hard copy or on a computer) where you'll be able find it easily. Also look for reports or project descriptions in the same general area as your own project. Detailed specifications about most component parts and devices are available on company Web sites. Over the past several years, these electronic databases have largely supplanted printed catalogs as the primary source of information for design engineers.

Perusing advertisements in trade magazines and journals can be a good way to learn what types of products are available that might be relevant to your project. Each field of engineering has many such publications. Examples include *Biomedical Products, Compliance Engineering, Computer Design, Electronic Component News, Electronic Design, Industrial Product Bulletin, Machine Design, Manufacturing Engineering, Mechanical Engineering*, and *OEM Technology News* (OEM: "original equipment manufacturer"). A good directory of technical magazines can be found at *www.techexpo.com/toc/tech_mag.html*.

### 2.4.3    Choose a Design Strategy

The next step in the design process of Figure 2.6 involves the selection of a strategy for meeting design objectives. At this stage of the project, the engineer (or, more likely, the design team) might decide whether the design will involve an electrical, mechanical, or software solution, and whether the product will be designed from the ground up or synthesized from off-the-shelf components. If the system is complex, it should be broken up into simpler, smaller pieces that can be designed independently and later interconnected to form the complete product. These subsections, or modules, should be designed so that they can be tested individually before the entire system is assembled.

Subdividing a large job into several more manageable tasks simplifies synthesis, testing, and evaluation. In a team design effort, the modular approach is essential. The various components of an automobile, for example, such as the engine, cooling system, electrical system, suspension, braking system, chassis, and drive train, are each designed and tested individually before the entire automobile is assembled. This same modular strategy applies to large software systems, where each section should be designed as an

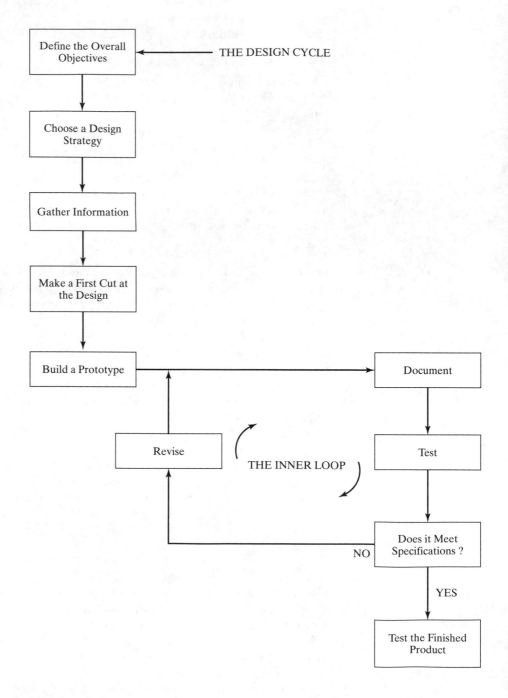

**Figure 2.6.** The design cycle.

independently testable module. In all cases, however, individual modules must always be designed with the overall design strategy in mind. Otherwise, the end product may consist of a collection of working but incoherent modules. In much of modular design, the integrated final product is more than the sum of its individual parts.

The design strategy also should consider similar efforts that may have been attempted in the past. Does a new technology exist that will be useful for the design? Perhaps a partial solution is already available in commercial form. For example, suppose that your design solution includes the amplification of voice or music. Many inexpensive off-the-shelf kits exist for constructing board-level amplifiers (see, for example, *www.rainbowkits.com*,) so that designing your own amplifier from raw electronic parts would be an unnecessary duplication of effort. The wise engineer makes use of existing products and components to simplify the design task. There is no shame in using off-the-shelf ingredients or subsystems if they can help achieve your design objectives more quickly and inexpensively. Typically, labor is the most expensive part of any development effort, so it's often cheaper (and sometimes more reliable) to buy something ready-made than it is to design it from scratch. Imagine how needlessly complex the task would be of designing a desktop computer without making use of the disk drives, memory chips, power supplies, monitors, and central processors available from other vendors. One caution: Be certain that using another company's product does not create patent infringement problems if your product is destined for commercial sale.

### 2.4.4    Make a First Cut at the Design

Once a design strategy has been identified, it's time to make an initial attempt, or "first cut" at the design. If the product is to be a physical entity, a layout or engineering drawing is helpful at this stage. Tentative values of dimensions, construction materials, part numbers, electronic component values, and other relevant parameters are specified. This step typically involves rough approximations and gross estimates. Its primary purpose is to determine whether or not the design approach has a chance of working, and it should result in a rough, tentative prototype for the system or each of its subsections. If the object of the design effort is a software product, its outer shell, overall structure, and user interface are laid out as part of the first-cut attempt. If the design involves a system (a manufacturing procedure, for example) than its overall flowchart should be specified at this time.

### 2.4.5    Build, Document, and Test

After the design team has reached a consensus on a first cut, a facsimile of the product is built as a working prototype. The typical first prototype is destined to be revised many times before the design cycle has been completed. This first prototype should be functional, but need not be visually attractive. Its primary purpose is to provide a starting point for evaluation and testing. For example, if the product is software, the prototype might consist of the main calculation sections without the fancy graphical interfaces that the user will expect in the final version. If the product is mechanical, it can be built up as a mock-up from easy-to-modify materials, such as wood or pre-punched metal strapping. For instance, a prototype for a new washing machine mechanism might be built inside an open wooden box made from framing lumber and hinged plywood. Such a structure would permit easy access to the inner machinery during testing, but obviously would never make it to the sales floor. If the product is electrical, its prototype could be wired on a temporary circuit breadboard. A new digital clock, for example, might be built in this way so that the designer could have easy access to its timing and display signals for test purposes. This arrangement would allow for circuit revisions before final construction and packaging.

During the prototyping phase of design, computer simulation tools such as AutoCAD™, ProENGINEER™, Solidworks™, PSPICE™, Electronics Workbench™, or Simulink™ can save time and expense by allowing you to predict performance before actual construction of the prototype. These software packages can help you identify hidden flaws before the product is built and give you some indication of the success of your design approach. However, computer simulations should never be used as a substitute for actual physical testing unless the product is very similar to one you've designed before. "Glitches," "bugs," and other anomalies caused by physical effects not modeled by the simulator have a nasty habit of appearing when a new product is tested. Despite the usefulness of computer-aided design tools and simulators, there is simply no substitute for constructing and testing a real physical prototype when the design is the first of its kind.

Note that documentation is part of the inner loop of the design cycle of Figure 2.6. The typical engineer faces considerable temptation to leave documentation to the very end of the design process. Pressed with deadlines and project milestones, many inexperienced engineers think of documentation as an annoying intrusion rather than as an integral part of the design process. After working diligently on a design project, the unseasoned engineer then faces with panic the reality of having to write documentation. ("Now I have to write this up?") Documentation added as an afterthought often is incomplete or substandard, because most of the relevant facts and steps have been forgotten by the time the writing takes place. Haphazard, after-the-fact documentation is the cause of many design failures. Many products, developed at great cost but delivered with pathetic documentation, have found their way to the trash heap of engineering failures because no one could figure out how to use or repair the product. Poor documentation also leads to subsequent duplication of effort and reinvention of the wheel, because no one can remember or interpret the results of previous work.

A good engineer recognizes that documentation is absolutely critical to every step of the design process. He or she will plan for it from the very beginning, keeping careful records of everything from initial feasibility studies to final manufacturing specifications. As the design progresses, it's a good idea to write everything down, even if it seems unimportant at the time. Information should be written in such a way that another engineer with the same technical background as your own could pick up your work at any time simply by reading your documentation. Careful documentation also will aid in writing product literature and technical manuals should the product be destined for commercial sale. Good documentation provides the engineer with a running record of the design history and the answers to key questions that were asked along the way. It provides vital background information for patent applications, product revisions, and redesign efforts, and it serves as insurance in cases of product liability. Above all, documentation is part of an engineer's professional responsibility. Its importance to engineering design cannot be overemphasized.

### 2.4.6   Revise and Revise Again

One of the characteristics of design that distinguishes it from reproduction is that the finished product may be totally different from what was envisioned at the beginning of the design cycle. Elements of the system may fail during testing, forcing the engineer to rethink the design strategy. The design process may lead the engineer down an unexpected path or into new territory. A good engineer will review the status of a product many times, proceeding through many revisions until the product meets its specifications. In truth, this revision process constitutes the principal work of the engineer. An experienced engineer recognizes it as a normal part of the design process and does not become discouraged when something fails on the first or second try. The revision cycle may require many iterations before success is achieved.

### 2.4.7    Thoroughly Test the Finished Product

As the design process converges on a probable solution, the product should be thoroughly tested and debugged. Performance should be assessed from many points of view, and the design should be modified if problems are identified at any stage. If the product is a physical entity, the effects of temperature, humidity, loading, and other environmental factors, as well as the effects of repeated and prolonged use, all must be taken into account. A physical product should be subjected to a "burn-in" (extended use) period to help identify latent defects that might cause the product to fail in the field. The human response to the product also should be assessed. No two people are exactly alike, and exposing the product to many different individuals will help identify problems that may not have been apparent during the development phase, when the product was examined by the design team only. Only after a comprehensive test period is the product ready to be put into actual service. Nothing will kill a new product faster than a few instances of malfunctioning in the field.

Like their physical counterparts, software products should be tested by a variety of different users who can discover hidden bugs. One of the characteristics of software is that different individuals will exercise the product in very different ways. Hence, extensive testing by a multitude of users is necessary if all software bugs are to be discovered. Commercial software is sometimes released to a control group of customers before widespread distribution. This control group understands that bugs may exist in the preliminary version of the software and are usually given incentives (reduced cost or a jump on competitors) to serve as real-world testers. A software trial of this type is sometimes called a *beta test* in the software industry.

---

**PROFESSIONAL SUCCESS:    HOW TO TELL A GOOD ENGINEER FROM A BAD ENGINEER**

As you pursue your engineering career, you will encounter many colleagues. Some will be good engineers, and others will be bad engineers. In your quest to identify and emulate only good engineers, you should learn the differences between the two. The following list highlights the traits of both types of engineers:

**A GOOD ENGINEER**

- Listens to new ideas with an open mind.
- Considers a variety of solution methodologies before choosing a design approach.
- Does not consider a project complete at the first sign of success, but insists on testing and retesting.
- Is never content to arrive at a set of design parameters solely by trial and error.
- Uses phrases such as, "I need to understand why," and "Let's consider all possibilities."

**A BAD ENGINEER**

- Thinks he/she has all the answers; seldom listens to the ideas of others.
- Has tunnel vision; pursues with intensity only the first design approach that comes to mind.
- Ships the product out the door without thorough testing.
- Uses phrases such as, "good enough", and "I don't understand why it won't work. So-and-so did it this way."
- Equates pure trial and error with engineering design.

**PRACTICE!**

1.   Without looking at Figure 2.6, draw a diagram of the design cycle that includes the following steps: define, gather, choose, first cut, build, document, test, revise.

2.   Suppose that you are asked to build a recumbent bicycle (a low-profile bicycle in which the pedals are in front of the seat). Make a list of the various ways in which you might gather information as part of the design cycle.

3.   Draw a modified design cycle, similar to Figure 2.6, that includes feedback from a test group of individual users.

4.   Define the design strategy that might have gone into the invention of the first personal computer. Imagine yourself before the days of small hard disk drives, graphical drivers for monitors, compact disks, and inexpensive memory chips. (In the early days of computers, random-access memory chips were one of the most expensive components of the PC.).

5.   Describe the various elements of the design cycle as they might have applied to the development of the ball point pen.

6.   Write a chronicle of the design cycle applied to a simple medieval catapult.

## 2.5   A DESIGN EXAMPLE

The typical engineering design cycle is made up of the following seven steps; 1) Define the objectives; 2) gather information; 3) choose a strategy; 4) make a first cut; 5) build, document, test; 6) revise (and revise again); and 7) test. Steps 5 and 6 form an iterative loop that may be repeated many times before the design effort is deemed successful. "Success" in this case is measured by how well the finished product meets the desired specifications. Imagine that your professor has announced a design competition as part of an introductory engineering course. The rules of the contest are outlined in the following flyer.

### COLLEGE OF ENGINEERING PEAK-PERFORMANCE DESIGN COMPETITION

**OBJECTIVE**

The goal of the competition is to design and construct a vehicle that can climb a ramp under its own power, stop at the top of the ramp, and sustain its position against an opposing vehicle coming up the other side of the ramp. The illustration in Figure 2.7 shows the approximate dimensions of the ramp. The 30-cm width of the carpet-covered track may vary by ±0.5 cm as the vehicle travels from the bottom to the top of the ramp. A vehicle is considered to be on "top of the hill" if its body, plus any extensions, strings, or jettisoned objects, lie completely within the two 120-cm lines after a 15-second time interval.

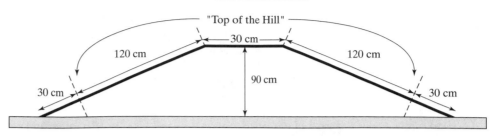

PEAK PERFORMANCE DESIGN COMPETITION
RAMP SPECIFICATIONS

**Figure 2.7.**  Ramp specifications for the Peak Performance Design Competition.

## VEHICLE SPECIFICATIONS

1. The vehicle must be autonomous. No remote power, control wires, or remote-control links are allowed.

2. The vehicle's exterior dimensions at the start of each run must not extend beyond the sides of an imaginary 30-cm cube. A device, such as a ram, may extend beyond this limit once activated, but cannot be activated before the start of the run.

3. The vehicle must be started by an activation device (e.g., switch or mechanical release) on the vehicle. Team members may not activate any device before the start. Vehicles cannot have their motors running before the start and dropped after the start.

4. The vehicle can be powered by the following energy sources only:

   - One battery of up to 9 volts
   - Rubber bands (4 mm × 10 cm maximum size in their unstretched state)
   - Mouse-traps (spring size 3 cm × 6 cm maximum)

5. The vehicle's weight, including batteries, must not exceed 2 kg.

6. The vehicle must not use chemicals or dangerous substances. No rocket-type devices, $CO_2$ propulsion devices, or chemical reactions are allowed. No mercury switches are permitted. (Mercury is a toxic substance, and a risk exists that a mercury switch will break during the competition.)

7. The vehicle must not be anchored to the ramp in any way before the start. At the end of the run, the vehicle and all its parts, including jettisoned objects, extensions, etc., must lie completely within the top of the hill and the 30-cm track width.

8. The vehicle must run within the 30-cm-wide, carpet-covered track. The vehicle may not run on top of the guide rails.

9. The vehicle must compete in six 15-second runs against opposing vehicles. The vehicles with the most wins after six runs will be selected for the Grand Finale. The latter will determine the winner of the competition. Modifications to the vehicle are permitted between (but not during) runs.

### 2.5.1   Applying Design Principles to the Design Competition

Let's now examine the Design Competition problem in the context of the seven steps of the design cycle. Remember that the problem can be addressed in many different ways. Some design solutions, however, will work better than others.

### 2.5.2   Define the Overall Objectives

When faced with the task of designing something, an inexperienced engineer is tempted to begin with construction right away without taking the time for a careful planning stage. Building things is fun and satisfying, while estimating, sketching, calculating, simulating, and checking design parameters seem less glamorous. It's important, however, to begin any project by taking time to define its objectives. In this case, your objectives might take the following form:

1.  *Design for speed.* The fastest vehicle will not necessarily be the winner, but in order to win, your vehicle must reach and maintain the center line well before the 15-second time limit. Otherwise, it may be blocked from reaching the top of the ramp by the other vehicle.

2.  *Design for defensive and offensive strategies.* Not only must your vehicle reach the top of the ramp and stop on its own, but it also must maintain its position as your opponent tries to do the same. Although offensive and defensive strategies are not necessarily mutually exclusive, you've decided (somewhat arbitrarily) that defense will be given a higher priority than offense. You may need to modify this choice if tests show it to be infeasible.

3.  *Design for easy changes.* The rules state that modifications to the vehicle are permitted between runs. During the contest, you may see things on other vehicles that will prompt you to make changes to your own vehicle. Similarly, you may choose to modify your vehicle in mid-competition should any of its features make it needlessly vulnerable. Adopting an easy-to-change construction strategy will facilitate on-the-fly changes to your vehicle. The likely trade-off in choosing this approach is that your car will be less durable and more likely to suffer a disabling failure.

4.  *Design for durability.* The vehicle must endure six, and possibly more, trips up the contest ramp. Opposing vehicles and accidents can damage a fragile design. You must weigh the issue of durability against your desire to produce a vehicle that's flexible and easy to modify.

5.  *Design for simplicity.* By keeping your design simple, you will be able to repair your vehicle quickly and easily. An intricate design might provide more performance features, but it also will be more prone to breakdowns and will be more difficult to repair.

Note that goals (1) through (5) are not independent of one another. For example, designing for easy changes may conflict with building a durable vehicle. Designing for both offensive and defensive strategies will lead to a more complicated vehicle that is harder to repair. Engineers typically face such trade-offs when making design decisions. Deciding which pathway to take requires experience and practice, but making any decision at all means that you've begun the design process.

### 2.5.3    Choose a Design Strategy

Many different design strategies will lead to a vehicle capable of competing. Building a *winning* design, however, requires making the right choices at each step in the design process. How can you know ahead of time what the right choices will be? In truth, you cannot, especially if you have never built such a vehicle before. You can only make educated guesses based on your experience and intuition. You test and retest your design choices, making changes along the way if they increase your vehicle's performance. This process of *iteration* is a crucial part of the design process. Iteration refers to the process of testing, making changes, and then retesting to observe results. Good engineering requires many iterations, trials, and demonstrations of performance before a design effort is completed. In the world of engineering, the first cut of a design seldom resembles the finished product.

In the case of the Design Competition described here, the rules provide for many alternatives in vehicle design. Regardless of the details of the design, however, all vehicles must have the same basic components: *energy source, propulsion mechanism, stopping mechanism,* and *starting device.* Although not required, a defense mechanism that prevents an opponent from pushing the vehicle back down (or off) the ramp will increase your chances of winning. After some discussion with your teammate, you develop the *choice map* shown in Figure 2.8.

This diagram displays some of the many design choices available to you and the consequences of making these design choices. For example, choosing battery power for propulsion constrains the stopping mechanism to one of the choices listed in the third

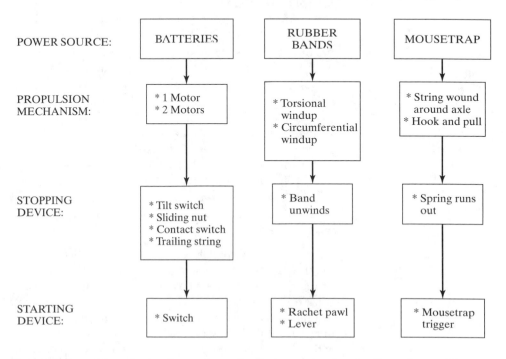

**Figure 2.8.**    Choice map that outlines the decision tree for the first phase of the design process.

box down. Although the diagram does not provide an exhaustive list of possible design choices, it serves as an excellent starting point for your design attempt. As your teammate points out, "We have to start somewhere." You remark in return, "Let's begin with our best guess as to what will work." The choice map of Figure 2.8 includes some of the following elements:

*Energy Source*   According to the rules of the competition, you may power your vehicle from standard nine-volt (9V) batteries, rubber bands, or mousetrap springs. Batteries are attractive because they require no winding or preparation other than periodic replacement. They will, however, be more expensive than the other two alternatives. Rubber bands will require much less frequent replacement, but will store the least amount of energy among the three choices. Like a rubber band, a mousetrap needs no frequent replacing. It stores more energy than a rubber band, but because of its physical form, it offers the fewest options for harnessing its stored energy.

*Propulsion Mechanism*   Your choice of the propulsion device, or *prime mover*, will depend entirely on your choice of energy source. If a battery is used as the energy source, an electric motor seems the obvious choice for turning the vehicle's wheels. Rubber bands can be stretched to provide linear motion or twisted for torsional energy storage that turns an axle or power shaft. Alternatively, a rubber band can be stretched around a shaft or spool like a fishing reel and used to propel the vehicle's wheels. A mousetrap can provide only one kind of motion. When released, its bale will retract in an arc, as depicted in Figure 2.9. This motion can be harnessed and used to propel the vehicle.

*Stopping Mechanism*   Your vehicle must stop when it arrives at the top of the ramp. This requirement can be met by interrupting propulsion power precisely at the right moment and relying on a combination of gravity and friction to stop the vehicle. A braking device to augment these forces also might be considered. If the vehicle is powered by batteries, there are many possibilities for a device to interrupt the flow of power to the vehicle. A simple tilt switch that disconnects the battery when the vehicle is level, but connects the battery when the vehicle is on a slope, will certainly do the job. For safety reasons, however, the rules prohibit the use of mercury switches (elemental mercury is toxic to humans), hence any tilt switch used in the vehicle must be of your own design. A metal ball bearing that rolls inside a small cage and makes contact with two electrodes, as illustrated in Figure 2.10, might serve as a suitable tilt switch. Other choices might be a spring loaded contact switch, such as the one shown in Figure 2.11, or a system that cuts off power to the wheels after the car has traveled a preset distance as measured by wheel rotations. This scheme will work well only if the wheels do not slip.

One interesting alternative to a mechanical switch would be to use an electronic timer circuit that shuts off power from the battery after a precise time interval. Through trial and error, you could set the elapsed time to just the right value so that the vehicle stops at the top of the ramp. One problem inherent with this open-loop timing system is that the vehicle does not actually sense its own arrival at the top of the hill, but rather infers it by precise timing. Because the speed of the vehicle may decrease with each successive run as battery energy is depleted, this timing scheme might cause problems. On the other hand, it is likely to be more reliable than solutions that involve mechanical parts.

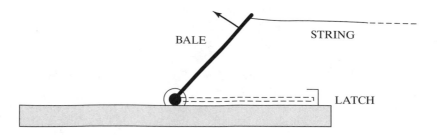

**Figure 2.9.**    Harnessing the stored mechanical energy of a mousetrap. The bale retracts in an arc when the mousetrap is released.

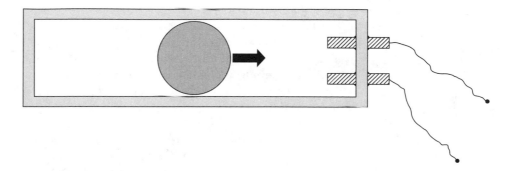

**Figure 2.10.**    Tilt switch made from a small enclosure, a ball bearing, and two contact points.

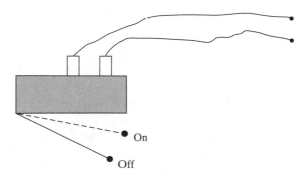

**Figure 2.11.**    Spring-loaded contact switch.

If a rubber band or mousetrap is chosen as the power source, then stopping the vehicle will require something other than an electrical switch. One crude way of stopping a vehicle propelled by mechanical energy storage is simply to allow the primary power source to run out (e.g., by allowing the rubber band to completely unwind). However crude this method, it is reliable because power input to the vehicle will *always* cease when the source of stored energy has been depleted.

***Starting Device***    If the vehicle is powered by a battery, then an electrical switch becomes the most feasible starting device. A rubber-band power source will require a mechanical device such as a pin or trip lever to initiate power flow to the wheels. A mousetrap can make use of its built-in trigger mechanism or any other starting mechanism that you can devise.

### 2.5.4    Make a First Cut at the Design

The first design iteration begins with rough estimations of the dimensions, parameters, and components of the vehicle to make sure that the design is technically feasible. After discussing the long list of design choices, you and your teammate decide upon a battery-powered vehicle. This decision makes available the many choices for a stopping device and power train. You feel that this design flexibility far outweighs the advantages of mechanical propulsion schemes. You decide upon a defensive strategy and agree to build a slower-moving, wedge-shaped vehicle driven by a small electric motor. The advantage of this design strategy is that the motor can be connected to the wheels using a small gear ratio, thereby providing higher torque at the wheels and a mechanical advantage that would be unavailable to a very fast vehicle. You plan to use plastic gears and axles purchased from an on-line Web site. The gear box will reduce the speed of the wheels relative to the motor shaft speed, providing added mechanical torque that will significantly increase the force available to push the opposing vehicle off the ramp. Because your vehicle will be slower than the others, it may not reach the top of the ramp first, but its wedge-shaped design will help to dislodge the front of any opposing vehicle that arrives first at the top of the hill. If your vehicle should happen to arrive first at the top of the hill, your car's defensive wedge shape will cause your opponent's car to ride over your car's body, allowing you to maintain your place at the top of the ramp.

You decide to use one motor with a single driven axle attached to both rear wheels. An alternative strategy would be to drive each rear wheel separately, thereby allowing the driven wheels to turn at different speeds. Such *differential* capability is essential for vehicles that travel curved paths, but in this case the vehicle must travel along a straight path only. By driving the wheels from a common shaft, you will reduce slippage, because both wheels will have to lose traction before forward motion is impaired. You briefly consider front-wheel drive, because you assume from hearing many car advertisements that front-wheel drive is superior to rear-wheel drive. Your teammate is quick to point out, however, that the advantage of front-wheel drive lies in its ability to help the car negotiate curves. Despite the media-driven message of "better traction," the advantages of front-wheel drive have nothing to do with your application. In fact, front-wheel drive may be a disadvantage to your wedge-shaped design, because it may cause your car to flip over forwards if another car travels on top of it (Figure 2.12). Your teammate draws the sketch of Figure 2.13 to illustrate this scenario. You abandon the idea of front-wheel drive.

### 2.5.5    Build, Document, Test, and Revise

A rough preliminary sketch of your car is shown in Figure 2.13. You've entered this sketch into a notebook in which you've been recording all information relevant to the project. Included in your notebook are design calculations, parts lists, and sketches of various pieces of the car. Shown in Figure 2.13 are the car's wedge-shaped design, a single drive shaft driven by a motor, belt, and pulleys, and a single switch to turn off the motor when the vehicle arrives at the top of the hill. Your design concept represents a trade-off between several competing possibilities, but you and your teammate have decided that the car's electric motor drive and defensive shape have the best chance of winning the competition.

REAR WHEEL DRIVE:

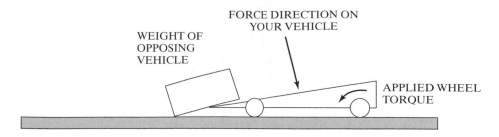

FRONT WHEEL DRIVE:

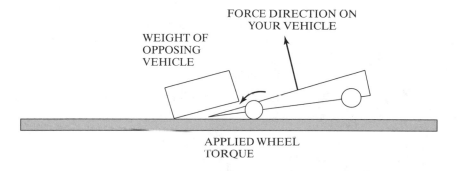

***Figure 2.12.***    Front- and rear-wheel drive options for a moving wedge vehicle.

The sketch in Figure 2.13 represents a beginning, but it is not the finished product. You still have many hurdles to overcome before your vehicle will be ready to compete. The next step in the design process involves building and testing a "first-cut" prototype. To help you in this phase of the design process, your professor has built a test ramp available to all contestants. You begin by constructing a chassis shell in the form of a wedge, without a motor drive or stopping mechanism.

You run your wedge-shaped vehicle up the ramp by hand. You soon discover that the bottom of the vehicle hits the ramp at the top of the hill, as depicted in Figure 2.14(a). The change in angle of the ramp is large, and all four wheels do not always maintain contact with the ramp surface. You discuss several solutions to this problem with your teammate. One solution would be to increase the size of the wheels, as shown in Figure 2.14(b). This change would decrease the mechanical advantage between the motor and the wheels, requiring you to recalculate the torque required from the motor. Another solution would be to make the vehicle shorter, as in Figure 2.14(c), but you realize that this solution would lead to a steeper angle for the wedge shape of the vehicle and reduce its effectiveness as a defensive strategy. (The sharper the wedge, the more capable the car will be of wedging itself under opposing vehicles. However, the largest thickness of the vehicle must stay the same to leave space for the motor and gear box.)

## 2.5.6   Revise Again

Your teammate suggests keeping the wheels and shape of the wedge the same and simply moving the rear wheels forward, as depicted in Figure 2.15. You rebuild the vehicle by moving the rear shaft mount forward, and you test the vehicle again. The redesigned vehicle no longer bottoms out on the ramp, and you claim success. Your professor sees your design changes and suggests that you test your vehicle under more realistic conditions. For example, what will happen when another vehicle rides over the top of your wedge-shaped body? You proceed to simulate such an event by placing a weight at various positions on the top of the car. The results of these additional tests suggest that moving the wheel location may not be the best solution to your problem. When you move the rear wheels forward, you change the base of support for the car's center of gravity. You discover that if an opposing vehicle rides over the top of your car, the net center of gravity moves toward the rear, eventually causing your car to topple backwards, as depicted in Figure 2.16.

### 2.5.7   Reality Check

These discoveries and setbacks may seem discouraging, but they are a normal part of the design process. Some things work the first time, while others do not. By observing and learning from failure and by building, testing, revising, and retesting, you can converge on the best solution that will meet your needs.

### 2.5.8   More Revisions

After some thought, you decide that increasing the size of the wheels may be the best option after all. Your teammate points out that you can simply change the ratio of the

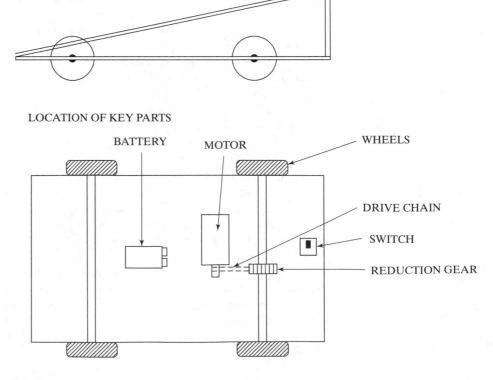

**Figure 2.13.**   Rough, preliminary sketch of a car for the Peak Performance Design Competition.

VEHICLE HITS RAMP

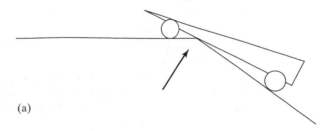

(a)

LARGER WHEELS

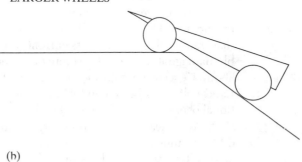

(b)

MAKE VEHICLE SHORTER

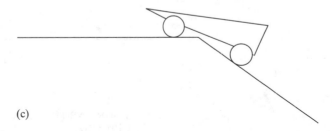

(c)

**Figure 2.14.** Vehicle at the top of the ramp. (a) Bottom of vehicle hits the ramp; (b) vehicle with larger wheels; (c) a shorter vehicle.

REAR WHEELS
MOVED FORWARD

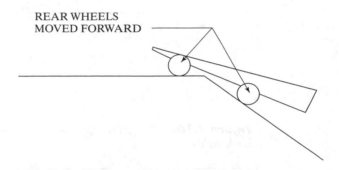

**Figure 2.15.** Moving the rear wheels forward.

gear box to preserve the net mechanical advantage between the motor and the wheels. This change will allow you to accommodate larger wheels. You buy some new wheels and try them with success. With the rear axle moved to its original location and the larger wheels in place, your car no longer bottoms out on the ramp.

You next consider the motor that will provide mechanical power to the drive shaft. Motors of all sizes and voltage ratings are available, including some alternating current (*ac*) motors, as well as direct current (*dc*) motors. Given that your car will be powered by batteries, your obvious choice is a dc motor. What voltage rating should you choose? You find no motors rated at 9 volts at the local hobby shop. "I don't think anyone makes 9-volt motors," says the salesperson behind the counter. You look up the Web pages of vendors,[2] scrutinize several catalogs, and find motors rated for 3, 6, 12, or 24 volts, but no 9-volt motors. Your professor explains that the rating of a motor specifies its operating voltage for continuous use. If a lower-than-rated voltage is used, the maximum torque available from the motor will be reduced. If the motor is connected to a higher-than-rated voltage, the excess current will heat the windings inside the motor and possibly damage it. During the competition, however, the motor will be energized only for about 15 seconds at a time. This interval may be short enough to allow a larger-than-rated voltage to be applied without damaging the motor. The feasibility of this intentional overloading must be verified by testing or by contacting the motor's manufacturer.

After hearing your professor's explanation, your team decides to purchase several different motors rated at 3 volts and 6 volts. You test each one by connecting it to a variable voltage supply and measure the current flow at several values of applied voltage.

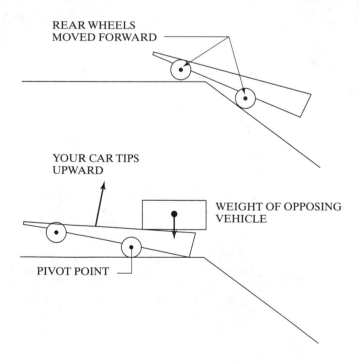

**Figure 2.16.**   Weight of opposing vehicle on top of rear end causes car to topple backwards.

---

[2] See, for example, *www.robotics.com* or *www.hobby-lobby.com*

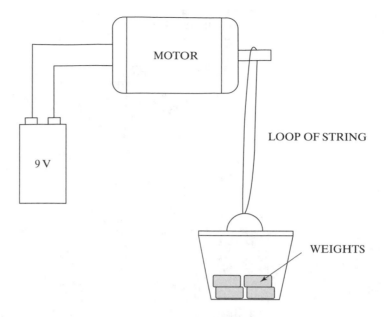

*MOTOR*

*LOOP OF STRING*

*9 V*

*WEIGHTS*

**Figure 2.17.**    Simple apparatus to measure motor torque.

You compute the power flow by multiplying the applied voltage by the current that actually flows to the motor. (Electrical power equals voltage times current.) You devise the apparatus shown in Figure 2.17 to measure the mechanical power delivered by the motor. Your contraption is a crude version of the industry-standard Prony brake used to measure motor torque.[3] The frictional rubbing of the weighted loop of string applies a mechanical load to the motor. You power each of your motors at the same voltage and add weights until the motor stalls. The motor that sustains the largest weight before stalling will have the highest torque. You check your mechanical loading measurements against your electrical measurements and use your data to determine which motor gives you the most mechanical torque when energized by a 9-volt power source.

As suggested by your professor, you and your teammate each keep a design notebook in which you record all your design decisions and the results of all your experiments and tests. Figure 2.18, for example, shows a page from your notebook in which you've recorded a list of the motors in your collection plus the results of the mechanical loading tests. You've also made a sketch of the loading apparatus of Figure 2.17 in your notebook. It's important to record the characteristics of the motors that you *don't* use, just in case your need to reconsider one of the rejected motors during a subsequent design revision. As your design proceeds, you record all calculations, specifications, and sketches pertinent to the drive train, including gear ratios, electrical power consumption, wheel diameters, weight of each part, and construction techniques. Your objective is to have a complete record of your design activities by the time you enter the vehicle competition.

---

[3] The Prony brake was invented in 1821 by Gaspard de Prony, a professor and examiner at the Ecole Polytechnique in France, as a way to measure the performance of machines and engines.

| MOTOR A: TC-254 | | | | MOTOR B: RS-257-234 | | |
| # weights | V(volts) | I (amps) | | # weights | V(volts) | I (amps) |
| 0 | 9.0 | 0.02 | | 0 | 9.0 | 0.05 |
| 1 | 8.9 | 0.05 | | 1 | 8.8 | 0.1 |
| 2 | 8.9 | 0.07 | | 2 | 8.7 | 0.17 |
| 3 | 8.8 | 0.1 | | 3 | 8.5 | 0.23 |
| 4 | 8.7 | 0.2 | | 4 | 8.0 | 0.3 |
| 5 | 8.0 | 0.5 | | 5 | 7.6 | 0.7 |
| | | | | | | |
| MOTOR C: BOD-37A | | | | MOTOR D: TOSH-7954 | | |
| # weights | V(volts) | I (amps) | | # weights | V(volts) | I (amps) |
| 0 | 9.0 | 0.01 | | 0 | 9.0 | 0.01 |
| 1 | 8.9 | 0.04 | | 1 | 8.9 | 0.11 |
| 2 | 8.9 | 0.05 | | 2 | 8.9 | 0.21 |
| 3 | 8.8 | 0.1 | | 3 | 8.8 | 0.31 |
| 4 | 8.7 | 0.25 | | 4 | 8.7 | 0.41 |
| 5 | 8.0 | 0.52 | | 5 | 8.7 | 0.51 |

***Figure 2.18.*** Page from lab notebook that documents mechanical loading tests on various motors.

## PRACTICE!

1. Make a two-column list that outlines the advantages of the various power sources for the Design Competition.
2. Make a list of additional propulsion mechanisms not mentioned in this section that could be used to drive a Design Competition vehicle.
3. Make a two-column list that outlines the advantages of using gravity and friction versus an applied brake as a stopping mechanism for the Design Competition.
4. Determine the minimum energy needed to lift a vehicle of the maximum allowed weight (2 kilograms) from the bottom of the ramp to the top.
5. How much electrical energy (in joules) is needed to exert one newton of force over a distance of 1 meter (m)?
6. How much electrical power (in watts) is needed to exert one newton of force on a body over a distance of 1 m for 10 seconds?
7. Determine the number of turns per centimeter (cm) of wheel diameter that will be required to move a vehicle from the bottom of the ramp to the top of the ramp in the Design Competition.
8. Determine the diameter of the wheels needed to move a vehicle from the bottom of the ramp to the top in the Design Competition with 50 turns of the drive axle.

## SUMMARY

The essential elements of the design process have been outlined at a very basic level. Design differs from analysis and reproduction because it involves multiple paths to solution, decision making, evaluation, revision, testing, and retesting. The *design cycle* is an important part of engineering problem solving, as are knowledge, experience, and intuition. *Documentation* is critical to the success of a product and should be an integral part of the design process.

The chapters that follow examine several aspects of design in more detail using the vehicle Design Competition as a continuing example. Some of the topics to be presented include teamwork, brainstorming, documentation, estimation, modeling, prototyping, and project organization. The use of computers is integrated into examples of these elements of engineering design.

## KEY TERMS

| | | |
|---|---|---|
| Design | Analysis | Reproduction |
| Design cycle | Decision making | Testing |
| Evaluation | Iteration | Revision |

## PROBLEMS

The following problem statements can be used to practice problem-solving skills and idea generation. Some of them involve paper designs, while others are suitable for actual fabrication and testing.

1. Develop a design concept for a mechanism that will allow hands-free operation of a standard, wired telephone. Outline its basic form, key features, proposed method of construction, and prototyping plan. Consider size, weight, shape, safety factors, and ease of use.

2. Develop a concept for a device that will allow hands-free use of a cellular telephone without the addition of additional accessories such as an earphone or remote microphone. Your design should be based solely on a mechanical solution.

3. Design a device for carrying a cellular telephone on a bicycle. Your contraption should enable the rider to converse while holding on to the handlebars with both hands.

4. Design a device for securing a coffee cup near the driver's seat of an automobile. The device should prevent the cup from spilling and should not interfere with the proper operation of the car. It should be universally adaptable to a wide variety of vehicles. Address safety and liability issues as part of your design.

5. Develop at least three design concepts for a nonlethal mousetrap. Is your device cost-effective compared with an ordinary, spring-bale mousetrap? Is your trap more humane, and is it worth any added cost?

6. Devise a concept for a device that can dig holes in the ground for the installation of fence posts. Sketch a prototype and outline a test plan for your design concept.

7. Design a device that will allow the inside and outside surfaces of windows to be cleaned from the inside only. Compare projected cost with that of a simple, hand-held, squeegee-type window cleaner.

8. Design a system for feeding a pet lizard automatically when the owner is out of town.

9. Develop a design concept for a spill-free coffee cup.

10. Develop a design concept for a coffee-cup to be used by the driver of an automobile. The cup should enable fully hands-free operation.

11. Design a device for carrying bricks to the top of a house for chimney repair. (The alternative is to carry them up a ladder by hand.)

12. Design at least three different systems for measuring the height of a tall building.

13. Design a system that will enable a self-propelled lawn mower to cut the grass in a yard unattended.

14. Design a system for minimizing the number of red lights encountered by cars traveling east and west through a major city. The system should not unduly impede north-south traffic flow.

15. Devise a method for managing the flow of two-way railroad traffic on a one-track system. The single line of track has a parallel spur track located every five kilometers. These parallel sections of track allow one train to wait as another passes by in the opposite direction.

16. Develop a concept for a public transportation system in which every traveler can ride a private vehicle on demand from any one station to another.

17. Design a transportation system based on locating free-use bicycles at strategic points around a large city.

18. Design a transportation system based on locating one-time, fee-for-use automobiles at strategic points around a large city.

19. Most large airports provide carts for travelers to transport luggage from baggage claim areas to taxi stands, transit stations, and parking lots. Develop a system for locating and reclaiming carts that have been taken from a central terminal station.

20. Develop a design concept for a system to measure the speed of a passing train.

21. Design a system that will assist flight attendants in feeding passengers on a medium-size airplane without blocking the single central aisle with food and beverage carts.

22. Design a system for automatically turning off a small electric baking oven when a cake inside is done.

23. Design a system that will help a person locate misplaced eyeglasses.

24. Develop a concept that will assist individuals in finding keys that have been misplaced around the house.

25. Design an electric switch that will turn off lights if the room is vacated, but not sooner than some user-specified time interval.

26. Design a device that will enable a quadriplegic to change the channels on a television set.

27. Devise a system for turning on security lights at dusk and turning them off at dawn. These lights are to be installed throughout a large factory. Your system should have a single master override for all lights.

28. Design a warning system for your home freezer that will alert you to abrupt changes in temperature.

29. Design a method or concept for dispensing adhesive tape (e.g., "Scotch" tape) in small, precut lengths.

30. Design a system that will automatically water houseplants when they are in need of moisture.

31. Design a system that will selectively water sections of a garden based on the moisture content of the soil at the location of each plant.

32. Design an irrigation system that will bring water from a nearby pond to your vegetable garden.

33. Develop a design concept for a system that will automatically pick apples from the trees of a commercial apple orchard.

34.  Design a system that can aerate the pond in a city park so that algae growth will not overtake other forms of wildlife.

35.  Design a device that will allow a one-armed individual to properly use dental floss.

36.  Design a system that will prevent the user of a battery charger from inserting the batteries the wrong way. The unit charges four AAA batteries simultaneously, but each battery must be inserted with its (+) end in the correct direction.

37.  Design a method for counting the number of people who attend a football game.

38.  Design a lamp post that will break away when struck by an errant automobile.

39.  Design a system for automatically steering an oil tanker along a desired compass heading.

40.  Design a weatherproof mail slot for a house that will keep cold air out but permit the insertion of mail from outside.

41.  Most small sailboats have a tiller (steering stick) in lieu of a steering wheel. Design a system for automatically steering a tiller-controlled sailboat on a course that lies along a user-specified compass direction.

42.  Design a system for automatically steering a tiller-controlled sailboat on a course that lies along a user-specified angle relative to the wind direction. Design for a simple boat that has no on-board electricity.

43.  One of the problems with recycling of postconsumer waste is the sorting of materials. Consumers and homeowners cannot always be relied upon to sort correctly, yet one erroneously placed container can ruin a batch of recycled material. At the present time, most municipalities resort to manual labor to sort recyclable materials. Devise a concept that will sort metal cans, plastic bottles, and plastic containers at a recycling plant. Develop a plan for modeling and testing your system.

44.  Design a kitchen device that will crush aluminum and steel cans in preparation for recycling. Such a device would be helpful for households that practice recycling but have limited storage space.

45.  Devise a system for painting car bodies automatically by robot. You must include a method for training the robot for each painting task.

46.  Devise a plan for a campus-wide information system that will allow any professor to access the grades of any student, while maintaining the privacy of the system to other users. A student also should be able to obtain his or her own grades, but not those of others.

47.  Design an electric pencil sharpener that will turn off when the pencil has been properly sharpened.

48.  Develop a design concept for a device that will turn on a car's windshield wipers automatically when rain falls. The wipers should come on only momentarily when rainfall is light but should be on all the time when the rainfall is moderate to heavy.

49.  Design an apparatus that will keep a telescope pointed at a distant star despite the rotation of the earth.

50.  Design a system for keeping a satellite's solar panels pointed at the sun.

51.  Devise a system for transferring personnel in flight from one airplane to another.

52.  Devise a system for automatically collecting tolls from cars traversing a major interstate highway. Note that many such systems exist throughout the United States. (See, for example, *www.mtafastlane.com* or *www.ezpass.com*.) This problem asks you to imagine (or find out) the details of how these systems work.

53.   Laser communication, or "laser-com," is a system by which digital data is sent from one location to another via a modulated laser beam. (See, for example, *www.microlink.hr/omnibeam.html*, or *www.usa.canon.com/indtech/broadcasteq/ canobeamdt50.html.*) Laser-com systems are used whenever hard connections via wires or fiber-optic cables are either too expensive or not possible. Laser-com links are much more difficult to tap for information than are other forms of wireless communication; hence, such systems are of obvious interest to the military for battlefield applications. One of the principal drawbacks of laser-com systems is the difficulty in maintaining beam alignment when the sender and receiver are moving. Design a system that will automatically direct the communication beams sent by two military vehicles to their respective receiving vehicles.

54.   Design a concept for a voice-synthesized prompting system that can provide cues for an individual who must take medication on a strict regimen. The device need not be pocket sized, but should be portable and battery operated.

55.   Design a system consisting of several panic buttons that will be installed at each of several workshop fabrication stations on a factory floor. Pressing any one of these buttons would activate a signal at a central control console and identify the location of the activated button. Ideally, voice communication over the system would be a desirable feature. One matter to consider is whether a wired or a wireless system is better.

56.   An elementary school teacher needs a calendar-teaching system to help students learn about dates, appointments, and scheduling events. Your professor has asked you to develop a design concept. The basic system should be a large pad over which a monthly calendar can be placed. The underlying pad should have touch-sensitive sensors that can detect a finger placed on each day block in the calendar. The entire unit should interface with a desktop computer which will run a question-and-answer game or program. The typical types of questions to be asked might include, "You have scheduled a dentist appointment 2 weeks from today. Point to the day on the calendar on which you should go to the dentist," or "Sara's birthday is on February 11. Point to that day on the calendar." An appropriate reward, either visual, auditory, or both, should be issued by the computer for correct answers. A nonintimidating signal should be issued for incorrect answers. Outline the key features of your system and devise a development plan.

*The following eight problems involve devices that will assist the professor running a Design Competition called "Peak Performance." In this competition, students are asked to build small, self-contained vehicles that travel up opposite sides of a trapezoidal ramp. The winner is the vehicle that lies closest to the top of the ramp after a 15-second time interval.*

57.   The rules of the Peak-Performance Design Competition state that each student-designed, autonomous vehicle be placed behind a starting line located 30 cm up the side of a 1.5-m. ramp. After the starting signal is given from the judge, contestants must release their vehicles, which then have 15 seconds to acquire and maintain a dominant position at the top of the ramp. Any vehicle that travels over the 30-cm starting line prior to the "go" signal loses the race. Currently, the starting sequence is initiated orally by the judge and timed by stopwatch. This system leads to great variability among judges, as many use different starting signals (e.g., "on your mark, get set, go!" or "one, two, three, go!"), and one may be lax in timing or checking for starting-line violations.

Design a system consisting of starting-line sensors, a start signal, a 15-second interval signal, and starting-line violation signals for each side of a double ramp.

The judge should have a button that initiates the start sequence. A series of periodic beeps that mimic the words, "ready, set, go!" should sound, with the final "go" being a loud and clearly distinguishable tone or buzzer. In addition, a green light or LED should illuminate when the "go" signal is sounded. The system should time for 15 seconds, then sound another tone or buzzer to indicate the end of the 15-second time interval. If a vehicle crosses the starting line prior to the "go" signal, a red light should go on for the violating vehicle's side of the ramp, and a special "violation" signal should be sounded to alert the judge.

58. Teams in the Peak Performance Design Competition are called to the floor when it is their turn to compete. After the initial call, each team has 3 minutes to arrive at its starting line. A team that does not arrive at the starting line after 3 minutes loses that run. Warnings are supposed to be given 2 minutes and 1 minute before the deadline. Traditionally, the announcer has called the teams and issued these warnings orally. With three races running simultaneously, all starting at different times, and with only some teams requiring the full 3 minutes to arrive, the proper issuing of these cues has been lax. The judges have asked that you design an automated system that will inform a given team how much of its 3-minute sequence has elapsed. The system must send an appropriate signal, oral, auditory, or visual, to only the affected team, and the timing sequence must be initiated from the judges' bench. As many as 80 teams may compete on a given day, and each is assigned one work table from a large array of 3 × 8-foot tables where the competition is held. Typically, the team being called is delayed, because it is repairing or modifying its vehicle.

    One of the key design issues is whether a wireless or wired system is better, given the logistic constraints of the competition environment. Because the event operates on a strict budget, final cost also is an important factor.

59. Vehicles entering the Peak Performance Design Competition must meet several requirements including a maximum battery-voltage limit. Each vehicle is checked once with a voltmeter at the start of the day by the head judge. Having a standardized voltage-checking device would shorten the time for voltage checking. Design a unit that has a rotary (or other type of) switch that can select a predetermined battery voltage. If the measured battery falls within the acceptable range, a green light should come on. If the voltage falls below or above the range, yellow and red lights, respectively, should come on.

60. Outline the design of a software tracking system for the Peak Performance Design Competition. Your system should enable the judges to automatically match up teams for matches, randomly at first, but by demonstrated ability thereafter (best teams against best teams; worst teams against worst teams). The program should display match sets before each round of the competition, and allow the recording judge to enter the result of each match after its winner has been determined. Assume a competition of up to 80 teams and six sets of matches between contenders.

61. Design a system for projecting matches on a display board so that the audience watching the Peak Performance Design Competition can keep track of who is racing who and the results of each match. Assume a competition of 20 teams with six sets of matches between contenders.

62. Design a system for detecting start-line violations for contenders in the Peak Performance Design Competition.

63. Design a system for determining the winner of each match in the Peak Performance Design Competition. Recall that the winner is the vehicle nearest the

center of the ramp after a 15-second time interval. Note that "nearest" can be an entirely subjective term; hence, your system for determining the winner must precisely define what "nearest" means.

64. Design a software system that will assist in registration for the Peak Performance Design Competition. Registration information includes name, address, and the school of the contenders. Registrants must pay a small fee, sign photograph permission and liability waiver forms, and receive an event T-shirt and assigned vehicle number. Several individuals work simultaneously on competition day to register participants as quickly as possible so that the event can begin on time.

65. A local company employs several workers who sort and package small (1 to 2 cm) parts in the 10 to 100-gram range. A typical operation might consist of putting 10 small parts in a polyethylene bag for subsequent packaging. As part of a course design project, your professor has asked you to design a mechanical sorting apparatus for dispensing these parts one at time so that the employees do not have to pick them up by hand. Develop an outline for how such a system might work, and draw a sketch of your proposed apparatus.

66. Ace Cleaning Services employs cognitively-challanged individuals performing a variety of cleaning tasks at four major buildings in the downtown area. Many of these workers have poor cognitive abilities and are unable to generalize the cleaning skills they have been taught. They cannot perform a supervisor-demonstrated office-cleaning task in a different office, even if a similar operation is involved. As a result, their cleaning job performance is often poor. Many of the employees rely on supervisor cues to repeatedly perform to acceptable standards the same task in different locations.

    A crew from Ace has been assigned to a 22-story office building with over 500,000 square feet of space that requires cleaning. Approximately 30 employees service this building between 6:00 a.m. and 10:00 a.m. with a staffing pattern of nine people. Workers are dispersed throughout the building to perform their daily cleaning routines. Supervisors are responsible for training and for checking that the work has been performed to acceptable standards. Many of the supervisors also perform direct labor, and hence cannot consistently provide prompting or cues to the rest of the employees all the time. A communication system is needed that can provide on-the-job cues to the cleaning staff. One system is needed for individuals who are literate and another for individuals who are not. Your task in this project is to design a modular system for either type of individual. The system must be designed so that additional units can be reproduced at low cost. As part of a project in your engineering design class, your professor has asked you to devise a system for providing recorded cues to Ace's employees. Develop a concept for the system, draw a sketch of one implementation, and outline how such a system might work.

67. A teacher at the nearby Carver School teaches a student who has severe developmental delays. This student is highly motivated by the Wheel of Fortune™ TV game show. The teacher currently has a 4-foot-diameter, wall-mounted, colorfully painted cardboard circle that spins and simulates the real game. Use of this wheel is supplemented by video clips of the game played on a VCR. The teacher would like a more elaborate, electronically-interfaced version of the game that enables the spinning dial to activate lights, voice, and the VCR clips. Despite the circus-like nature of this project, it is a top priority of the Carver school system. The customer is in need of an imaginative and creative response to this problem, and your engineering professor has asked you for ideas. Sketch several versions of the

system, highlighting the advantages and disadvantages of each approach. How would you test the success of your design?

68.  A teacher wants a clock system that can help students to learn the relationship between time displayed by digital clocks and time displayed by analog clocks. The system should have a console that contains a large analog clock face, as well as a digital clock display with large digits. In operation, the teacher will set either clock, then ask the student to set the other clock to the same time. If the student sets the time correctly, the unit should signal the student appropriately. If the student fails to set the time correctly, the unit should also issue an appropriate response. Outline the salient features of such a system.

69.  Your school has been asked by an individual confined to a wheelchair to build a small motorized flagpole that can be raised and lowered by pressing buttons. The person needs such a device to hold a bright orange flag to provide visibility outdoors while navigating busy city streets and sidewalks. The flag must be lowered when the individual enters buildings so that the pole does not interfere with doorways and low ceilings. Here is a copy of the letter received by your school:

*East Crescent Residence Facility*
*Eleven Hastings Drive*
*West Walworth, MA 02100*

Prof. Hugo Gomez
College of Engineering
Correll University
44 Hartford St.
Canton, MA 02215

May 18, 2001

RE: Retractable Flag for Wheelchair

Dear Professor Gomez,

I am the supervisor of a residence facility that services adults with special needs. One of our residents is confined to a wheelchair and spends a great deal of time traveling throughout the community, often on busy streets, in a motorized wheelchair. Although a flag on a long pole would increase her safety, she is reluctant to install one on her wheelchair, because it becomes a problem in restaurants, crowded stores, and on public buses. I was wondering if you might have some students who could design an electrically retractable flag, possibly with visual enhancement (e.g., a flashing light) that could be raised or lowered by the individual on demand. The flag deployment mechanism could operate from either the wheelchair's existing automobile-type storage battery or its own self-contained battery. If such a device is possible, could you give me a call? I would appreciate any assistance that you might be able to offer.

Sincerely yours,

Liz DeWalt
Director

Prof. Gomez has asked you to try to build such a device for Ms. DeWalt. How would you approach such a task? Such a seemingly simple device actually can be more complicated to design than you might think. Draw a sketch of the wheelchair device, then devise a design plan for building and testing several designs. Include a list of possible safety hazards to bystanders and the user. How can you include the user in the design process, and why is it advisable to do so? Also write a report of your preliminary findings for Prof. Gomez. Your design strategy should begin with a conceptual drawing of the device that you can send to Ms. DeWalt for comments. Generate a specification list and general drawing of the apparatus as well as a cover letter to Ms. DeWalt.

*Design Considerations:* One goal of your design might be to make a flag device that can be mounted on any wheelchair, not just on that of Ms. DeWalt's client. Because many wheelchairs are custom designed for the user, your device must be easily adaptable for mounting on different wheelchair styles. Not all wheelchairs are motorized, hence your device must operate from its own batteries to accommodate hand-pushed wheelchairs. Another consideration in favor of separate battery power is that motorized wheelchair manufacturers usually specify that no other electrical or electronic equipment be connected to the primary motor battery for reasons of safety, reliability, and power integrity.

One last consideration concerns the placement of the switch needed to activate the flag. Like the flag itself, the activation switch must easily attach to structural features of the wheelchair and must be within easy reach of the user. At the same time, it must not be so obtrusive that it distracts the user when not in use and must not hurt anyone.

70. Develop a design concept for a computer-interfaced electronic display board that can be placed in the lobby of an office building to display messages of the day, announce upcoming seminars, or indicate the location of special events. The objective of the problem is to use a matrix of addressable light-emitting diodes (LEDs) rather than a video display. The system should accept messages by wire from a remote site. One approach might be to design your display board system so that it is capable of independently connecting to a local-area computer network. Alternatively, you could build a separate remote device that could be connected to a desktop computer and then brought down to the display board to load in the data. These examples are suggestions only. In general, any means for getting data to the board is acceptable, but a separate computer (PC) cannot become a dedicated part of the finished display.

71. An engineer is interested in measuring the small-valued ac magnetic fields generated by power lines and appliances. You have been asked to design a battery-powered, hand-held instrument capable of measuring the magnitude of ac magnetic fields in the range 0.1 to 10 $\mu$T (microtesla) at frequencies of 50-60 Hz. Magnetic fields of this magnitude are very small and are difficult to measure accurately. For comparison, the earth's dc magnetic field is on the order of 50 $\mu$T, and the magnetic field inside a typical electric motor is on the order of 1 T.

Using your knowledge of physics, summarize the important features that such a device should have. Outline a design plan for its development and construction. You have several options for the primary sensor. For example, it may consist of a flat coil of wire of appropriate diameter and number of turns, or, alternatively, you might consider using a commercially available semiconductor sensor. Note that dc fields, such as those produced by the earth or any nearby permanent magnets, are not of interest. Hence, any signal produced by dc fields in your instrument should be filtered out. Ideally, your unit should have a digital or analog display device and should accommodate a remote probe if possible.

72. A friend of yours runs a residence home for individuals who are mentally and physically challenged. She would like a medicine dispenser for dispensing pills at specific times. The unit is to be carried by an individual and must have sufficient capacity to hold medication for at least 1 day. The unit should open a cassette or compartment and should emit an audible or visual signal when it dispenses medication. The unit must be easy to load and should be easily programmable by a residence-home supervisor. Your friend has asked you to assess the feasibility of developing such a unit. Get together with one or more other students. Discuss the feasibility of the idea and develop several design concepts for further evaluation.

73. A friend of yours is an enthusiast of remote-control model airplanes. One perennial problem with radio-controlled airplanes concerns the lack of knowledge about the flight direction and orientation of the airplane when it is far from the ground-based operator. When the airplane is too far away to be seen clearly, the operator loses the ability to correctly control its motion. Develop a design concept for a roll-, pitch-, and compass-heading indicator system that can be mounted on the model airplane and used to send the information via radio back to the operator's control console. Your system should sense the pitch and roll of the airplane over the range +90 degrees to −90 degrees and be able to withstand a full 360-degree roll or loop-de-loop.

74. Your family has asked you to design a remote readout system for a vacation home to be interrogated by a remote computer over a modem and telephone line. The unit in the vacation home should answer the phone after 10 rings, provide means for an entry password, and then provide the following information: inside and outside temperatures, presence of any running water in the house, presence of any loud noises or unusual motion, and status of alarm switches installed on doors and windows. Discuss the design specifications for such a unit and develop a block diagram for its design and implementation.

# 3

# Working in Teams

The engineer uses many technical skills when designing a device, product, or system. These skills represent the knowledge base acquired through years of study and on-the-job training. They are all but useless, however, if the engineer is unable to communicate and work with others. In addition to a strong set of technical skills, a good engineer must have nontechnical skills that include the ability to work in a team, generate new ideas, keep good documentation, and work within the framework of project management. Seldom does an engineer simply sit down and get right to work on the technical details of a project. Before beginning work in earnest, he or she spends much time planning, conducting feasibility studies, reviewing results of other projects, doing approximate calculations, interacting with other engineers, and defining the approach to the problem. This chapter introduces several teamwork skills that are essential elements of successful design.

## SECTIONS

- 3.1 Teamwork Skills
- 3.2 Brainstorming
- 3.3 Documentation: The Key to Project Success
- 3.4 Project Management: Keeping the Team on Track

## OBJECTIVES

*In this chapter, you will learn about*

- Teamwork as an essential element of engineering design.
- Brainstorming.
- Documentation and its vital role in the design process.
- Project management skills.

## 3.1 TEAMWORK SKILLS

The spirit of rugged individualism persists as a theme in books, movies, and television. The image of a lone hero striving for truth and justice against insurmountable odds appeals to our sense of adventure and daring. The dream of becoming the sole entrepreneur who endures economic and technical hardship to change the face of technology, to head a startup company of one that single-handedly takes on Microsoft or another large corporation, arouses our pioneering spirit. Yet, in the real world, engineers seldom work alone. Most engineering problems are interdisciplinary. True progress requires teamwork, cooperation, and the contributions of many individuals. This concept is easy to understand in the context of designing large structures, such as bridges, airports, or global computer networks. Teamwork is also critical to the design of complicated devices, such as automobiles, video players, medical implants, network routers, copy machines, or cellular telephones. These devices, and others like them, cannot be designed by one person alone. The great engineering accomplishments in space exploration, such as the Apollo moon landing, the International Space Station, and the Hubble space telescope, required hundreds (in some cases thousands) of engineers working with teams of physicists, chemists, astronomers, material scientists, medical specialists, mathematicians, and project managers. Teamwork is an important skill, and you must master it if you are to be a good engineer. Working in a team, as in Figure 3.1, requires that you speak clearly, write effectively, and have the ability to see another person's point of view. Each member of a team must understand how his or her task relates to the responsibilities of the team as a whole.

Many engineering firms offer team-building workshops as part of their employee training programs. Self-help books on the subject of teamwork abound. You'll have many opportunities to work as a team if you study engineering, and you should treat each one as a learning experience.

### 3.1.1   Effective Team Building

An effective team is one that works well together. It functions at its maximum potential when solving a design problem and thrives on the special capabilities of its individual members. One key characteristic of an effective team is a good supportive attitude among fellow teammates and team activities. Team morale and a sense of professionalism can be enhanced if team members agree upon some rules of behavior. The following set of guidelines illustrates one possible approach to building an effective design team.

### *1. Define Clear Roles*

Each team member should understand how he or she is to function within the team. The responsibilities of each individual should be defined *before* work begins on the project. Roles need not be mutually exclusive, but they should be defined so that all aspects of the design problem fall within the jurisdiction of at least one person. In that way, no task will "fall between the cracks" during the design process.

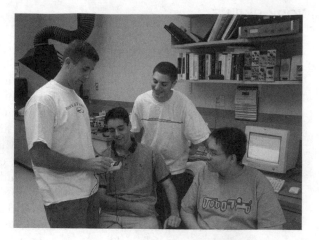

***Figure 3.1.*** Students working on a design project build strong team relationships.

### 2. Agree Upon Goals

Members of the team should agree upon the goals of the project. This consensus is not as easily achieved as you might think. One teammate may want to solve the problem using a traditional, time-tested approach, while another may want to attempt a far-out, esoteric path to success. Define a realistic set of goals at the outset. If the design process brings surprises, you can always redefine your goals midway through the project.

### 3. Define Procedures

Teammates should agree on a set of procedures for getting things done. Everything from documentation and the ordering of parts to communication with professors, clients, and customers should follow a predetermined procedure. In that way, misunderstandings about conduct can be greatly reduced.

### 4. Develop Effective Interpersonal Relationships

You must learn to work with everyone on your team, even with those individuals whom you may personally dislike. In the real world, a client will seldom care about any conflicts that occur behind the scenes. It's a sign of engineering professionalism to be able to rise above personality clashes as you concentrate on the job at hand. Be nice. Be professional. Forbid name calling, accusations, and assigning fault between team members.

### 5. Define Leadership Roles

Some teams work best when a single person emerges as a chosen leader. Other teams work better by consensus using distributed leadership or even no leadership at all. Regardless of your team's style, make sure that leadership roles are clearly defined and agreed upon at the start of a project.

**PROFESSIONAL SUCCESS:   YOU'RE THE TEAM LEADER, ONE TEAMMATE HAS DISAPPEARED, AND YOU'RE DOING ALL THE WORK**

It's impossible to get along with all people all the time. When you work closely with other individuals, personal conflicts are inevitable. At times, these disagreements occur because one team member has failed to meet his or her responsibilities. At other times, the conflict arises from fundamental differences in personal outlook or priorities. However complicated your team relationships may become, remember that your customer does not care about them. Your customer is interested in receiving a well-designed product that reflects your best engineering abilities. It's up to you to resolve team conflicts internally. This resolution may mean that some team members will do more work than others, even if they will not be rewarded for their extra efforts. A good leader understands this trade-off and devises a plan to work around an errant teammate. Such situations may seem frustrating and unfair, but they happen in the real engineering world all the time. Learning how to deal with them as a student is part of your engineering education.

**PRACTICE!**

1. Define the roles, goals, and procedures that might apply to the design and construction of an eight-lane highway system under a major metropolitan city (*www.bigdig.com*).

2. Define the roles, goals, and procedures for a team of software engineers developing a Web site for the sales catalog of a national hardware store (*www.truevalue.com*).

3. Define the roles for a team of electrical and mechanical engineers developing a solar-powered car (*www.solarcar.arizona.edu*).

4. Define the roles, goals, and procedures that might apply to a team of biomedical engineers developing an artificial heart for mechanical implantation inside human subjects (*www.abiomed.com*).

5. Show how the five elements of effective team building might apply to the functioning of a soccer team (*www.fysa.com*).

6. Define the roles, goals, and procedures that might apply to the design and construction of an international space station (*www.boeing.com/defense-space/space/space station*).

## 3.2   BRAINSTORMING

One obvious area where teamwork plays an important role is the generation of new ideas for problem solving. When engineers gather to solve problems, they often resort to a creative process called brainstorming. Brainstorming requires a spontaneous mode of thinking that frees the mind from traditional boundaries. All too often, we limit our problem-solving approach to obvious solutions that have worked in the past. Responsible engineering sometimes requires that we consider other design alternatives, including those previously untried. A good engineer will never settle on a solution just because it's the first one to come to mind. When engineers brainstorm, creativity proceeds spontaneously unfettered by concerns that an idea is "way out" or impractical. When the constraints of traditional paradigms are removed, new solutions often emerge. Hearing the ideas of your teammates can tap the ideas buried deep inside your brain. Promising, but different, ideas are discarded as unfeasible only after study, analysis, and comparison with competing ideas. Brainstorming allows the engineer to consider as many options as possible before choosing the final design path.

Brainstorming can be done informally, or it can follow one of several time-tested formal methods These formal methods are appropriate for large group settings where organization is needed to avoid chaos. Less formal brainstorming methods are reserved for groups of one to three people who wish to generate ideas. Although they differ in execution, formal and informal brainstorming methods share the same set of core principles. The primary goal is to foster the uninhibited free exchange of ideas by creating a friendly, nonjudgmental environment. Brainstorming is an art. It requires practice, but anyone who has an open mind and some imagination learn this important skill.

### 3.2.1 Ground Rules for Brainstorming

The ground rules for brainstorming are designed to create a friendly, nonthreatening environment that encourages the free flow of ideas. Although the specific rules may vary, depending on the procedures followed, the following list can serve as a guideline:

1. No holding back. Any idea may be brought to the floor at any time.
2. No boundaries. An idea is never too outrageous or "way out" to mention.
3. No criticizing. An idea may not be criticized until the final discussion phase.
4. No dismissing. An idea may not be discounted until after group discussion.
5. No limit. There is no such thing as having too many ideas: the more the better.
6. No restrictions. Participants may generate ideas from any field of expertise.
7. No shame. A participant should never feel embarrassed about bringing up what seems like a stupid idea.

When a brainstorming session is in progress, one person should act as the facilitator, and another should record everyone's ideas. It's also possible to use video or audio tape in lieu of a human secretary.

### 3.2.2 Formal Brainstorming Method

When a large group gets together to brainstorm, formal structure is helpful. Without such a structure, a flood of competing ideas, all brought to the floor simultaneously, can create chaos. Instead of thinking creatively, participants become confrontational as they strive to be heard and gain a voice in the conversation. With so many randomly competing opinions, each person's creative process is inhibited, and the brainstorming session becomes unproductive. This effect is sometimes called "idea chaos." Adding formal structure to the brainstorming session restricts the flow of ideas to a manageable rate without restricting the number of ideas generated. In fact, the addition of a formal structure in large group settings can enhance the brain's creative process by preventing aggressive individuals from dominating the conversation and by providing time for people to think.

Of the many formal brainstorming techniques that exist, the *idea trigger* method has been well tested and is used often by brainstorming specialists in large group settings. The idea trigger method is based on the work of psychologists[1] and has been shown to enhance the brain's creative process. It relies on a process of alternating     tension and relaxation that taps the brain's creative potential. By listening to the ideas of others, receiving the foreign stimulus of other people's spoken ideas, and being forced to respond with counter ideas, a participant's behavioral patterns, personality constraints, and narrow modes of thinking can momentarily be broken, allowing ideas hidden in the recesses of the brain to come to the foreground. A shy participant who is reluctant to offer seemingly silly ideas, for example, will be more willing to do so under the alternating tension and relaxation of the purge-trigger sequence.

The idea trigger method requires a leader, at least four participants, and a printed form such as the one shown in Figure 3.2. The procedure has three phases, as follows.[2]

*Phase 1: Idea Generation Phase*     The problem or design issue is summarized by the leader. Each person is given a blank copy of the form shown in Figure 3.2. Without talking, each participant writes down in rapid succession as many ideas or solutions as possible. These entries are placed under Column 1. Key words suffice; whole sentences are not necessary. During the idea-generation phase, participants open their minds, consider many alternatives, and do not worry if ideas seem too trivial or ridiculous. "Pie in the sky" radical, or impossible ideas should be included. In short, participants write down anything that comes to mind that may be relevant to the problem. The fact that ideas are written down silently removes the element of intimidation from the idea-generation process.

After the first 2 minutes of the session, the group takes a break and then attempts to write down additional ideas under Column 1 for another 60 seconds. This *tension and relaxation* sequence has been shown to enhance creativity. It helps to extract all ideas from the brain's subconscious memory, much like squeezing and releasing a sponge several times to extract all the water.

*Phase 2: Idea Trigger Phase*     After the idea generation phase, the leader calls upon all members of the group. Each participant takes a turn reading his or her entries from Column 1. As each person recites Column-1 entries, others silently cross out the duplicates on their own lists. Hearing the ideas of others will trigger new ideas, which each person should enter under Column 2 as soon as they emerge. This process is called *idea triggering*. Hearing the remarks of others while pausing from the act of speaking causes the hidden thoughts stored in the subconscious to surface. The purpose of the idea trigger phase is not to discount the ideas from Column 1, but rather to amplify them, modify them, or generate new ideas.

After all members have read their Column-1 entries and have completed their Column-2 entries, the idea trigger process is repeated again. This time, entries from Column 2 are read, and any new ideas triggered are entered under Column 3. The process is repeated, with entries added to Columns 4, 5, etc., until all ideas are exhausted. Complex problems may require as many as five rounds of idea trigger phase.

---

[1] G. H. Muller, *The Idea Trigger Session Primer*, Ann Arbor, MI: A.I.R. Foundation, 1973. S. F. Love, *Mastery and Management of Time*, Englewood Cliffs, NJ: Prentice-Hall, 1981.

[2] C. Lovas, *Integrating Design into the Engineering Curriculum*, Dallas, TX: Engineering Design Services Short Course and Workshop, October 1995.

IDEA TRIGGER SESSION

| COLUMN 1 | COLUMN 2 | COLUMN 3 | COLUMN 4 |
|----------|----------|----------|----------|
| 120 MINUTES | | | |
| | | | |
| 60 SECONDS | | | |

CONTRIBUTOR:

***Figure 3.2.*** Blank form for the idea trigger session.

The entries that appear under the second and third columns (and the fourth and fifth columns if the problem is complex) are usually the most creative. Such richness results from several factors. Often participants are secretly angered at having had their ideas stolen by another. This simple competitive pressure can propel a person toward new, unexplored territory. Conversely, seeing that one's ideas have not been duplicated by others can provide positive reinforcement, pushing the participant to come up with even better ideas. Some individuals may respond to their own unduplicated entries with a desire to produce more as a way of hoarding the good ideas. Yet others may subconsciously think that augmenting previously discussed ideas fosters group cooperation.

***Phase 3: Compilation Phase*** When the idea trigger phase has been completed, it's the job of the leader to compile everyone's sheets and make one master list of all the ideas that have been generated. The group then proceeds to discuss all ideas, discarding the ones that probably will not work, and deciding which of the remaining ideas are appropriate for further consideration and development.

**EXAMPLE 3.1:
A FORMAL
BRAIN-
STORMING
SESSION**

Let's illustrate the formal idea trigger method with an example. Four students, Tina, Juan, Paul, and Karin, are designing an entry for a design competition. The overall objective of the competition is to design a self-propelled vehicle that can climb a 1.5-meter ramp, stop at the top, and prevail over an opposing vehicle climbing up the ramp from the other side. The four students recently held a brainstorming session using the idea trigger method. They addressed all elements of the car design, including the issues of propulsion, offensive and defensive strategies, and a stopping mechanism. The following discussion chronicles their brainstorming session. Tina acted as the leader and timed the first 2 minutes, the break, and the subsequent 60-second idea generation phase. At the end of the phase, Juan's page looked like this:

| JUAN | IDEA GENERATION PHASE COLUMN 1 (2 MINUTES): |
|------|------|
| | Support structure = wood (easy to make) |
| | Use angle irons from Mechano™ |
| | Plastic body for lighter weight |
| | Zinc air batteries (lightweight) |
| | Wheels taken from my radio-controlled car |
| | Rubber band for chain drive |
| | Small car will be harder for opponent to deflect |
| | **1 MINUTE:** |
| | Ramming device |
| | Wedge-shaped body |

Juan read his entries. As Paul listened he crossed out his own duplicate entries. When Juan was finished, Paul's first column, including crossouts, looked like this:

| PAUL | IDEA GENERATION PHASE COLUMN 1 (2 MINUTES): |
|------|------|
| | ~~No heavy batteries (use zinc air)~~ |
| | Larger wheels for slower turning speed |
| | Gear box |
| | Higher torque (harder for opponent to push backwards) |
| | ~~Use plastic for body~~ |
| | Electronic timer for stopping mechanism |
| | Rechargeable batteries |
| | ~~Wedge-shaped design~~ |
| | **1 MINUTE:** |
| | ~~Buy wheels from hobby shop for radio-controlled car~~ |
| | Sense speed, determine distance traveled |
| | Aluminum frame |

Next Karin read those of her entries that had not been duplicated by Juan. As Paul listened an idea flashed into his head. *A threaded rod*, he thought. *We can make the drive shaft from a threaded rod.* Paul reasoned that they could make the drive shaft from a threaded rod and have it screw a sliding nut toward a cutoff switch. The method would not be foolproof, because slipping wheels could ruin the system's ability to track distance. It seemed worth discussing, though, so Paul wrote down "threaded rod" under his Column 2 entries.

When Karin heard Juan read his "ramming device" entry, it had made her think about using an ejected object as part of an offensive strategy. She wrote the words "ejected device" under her Column 2 entries. Tina reacted similarly to Juan's idea and wrote the words, "lob something on the track ahead of opposing car" under her Column 2 entries. The spoken trigger phase made its way around the group. When everyone had finished, Tina acting as leader, started the process again. This time, everyone read their Column 2 entries and wrote down new ideas under Column 3. As Karin read her entry about ejected devices from Column 2, Tina got another idea. The idea of an ejected object brought to her a fleeting image from the Herman Melville novel *Moby Dick.* She imagined a flying spear with a barbed tip shot ahead of the vehicle over the top of the hill. *After hitting the carpet in front of the opposing vehicle,* she thought, *the barbed tip will dig into the carpet, blocking the other car. This spear will be very difficult to dislodge.* Tina wrote down "harpoon" under her Column 3 entries.

The second idea-trigger round progressed, and Tina started a third. After about 45 minutes, the entire Phase 2 session was finished. Tina suggested a break so that she could compile everyone's lists of ideas. Her combined list of entries from everyone's three columns looked like this:

---

**COMPLETE LIST OF IDEAS FROM EVERYONE'S SHEETS**

---

**SHAPE:**

Small car = harder for opponent to deflect

Wedge-shaped vehicle having same width as track

Rolling can design

Snow-plow shaped wedge

**STRUCTURE:**

Support structure = wood (easy to make)

Aluminum frame

Plastic body for light weight.

Use angle irons from Mechano™

Hot-melt glue balsa wood

**POWER:**

Zinc air batteries (lightweight)

Rechargeable batteries

Change batteries after every run

Electronic timer for stopping mechanism

Microprocessor-controlled car with onboard sensors

Sense speed, determine distance traveled from microprocessor software

**PROPULSION:**

Wheels from radio-controlled car purchased at hobby shop

Large wheels

---

---

**COMPLETE LIST OF IDEAS FROM EVERYONE'S SHEETS**

**PROPULSION:**

Rubber band for chain drive

Plastic-linked chain from junked radio-controlled car chassis

Single large mousetrap with mechanical links

Wind-up large rubber band

**STRATEGIES:**

Ramming device

Flying barbed harpoon

Pickup arm

Throw jacks in front of oncoming opponent

Roll over opponent with large roller

---

After the break, Tina reconvened the team to discuss the list of ideas. They weeded out the ones that did not seem feasible and compared ideas that looked promising. They combined multiple ideas and converged on a slow-moving, wedge-shaped vehicle concept for the prototype stage. They also decided to try out Tina's offensive strategy of a flying harpoon designed to dig into the carpet and block the path of the opposing vehicle.

### 3.2.3 Informal Brainstorming

The formal brainstorming method discussed in the previous section requires organization and planning. In contrast, informal brainstorming can be done anywhere. As a technique for engineering design, informal brainstorming in a round table format is appropriate for small groups of people. Ideas are contributed in random order by any participant. The flow of ideas need not be logical, and new proposals can be offered whenever they come to mind. The previously introduced ground rules should still be enforced during an informal brainstorming session.

**EXAMPLE 3.2: INFORMAL BRAIN-STORMING**

The following example of an informal brainstorming session describes a conversation that might have taken place between two team members working on a design competition in which the objective is to design a battery-powered vehicle capable of climbing a ramp and stopping at the top. In this case, Tina and Juan discuss the problem of how to stop the vehicle when it arrives at the top of the ramp. Note the ebb and flow of ideas between the two students. They do not immediately fixate on the first idea that comes to mind, but instead allow the flow of ideas to lead them to the solution most likely to succeed.

*Tina*: "Let's use a switch to turn off the electric motor when the car gets to the top of the ramp."

*Juan*: "OK, we could." *He thought for a while.* "We could also modify the switch so that it trails behind and is springloaded into the closed position. Putting the car on a flat surface will press the lever arm, close the switch, and connect the battery to the motor. When the car reaches the top, the switch arm will stay on the slope, allowing the switch arm to spring downward. The switch will open, disconnect the battery from the motor, and stop the car." (Juan drew the sketch shown in Figure 3.3.)

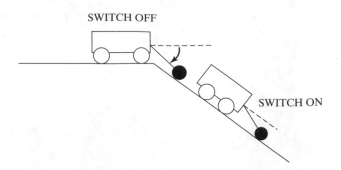

**Figure 3.3.** Juan's idea for a stopping switch.

*Tina*: "That *might* work, but maybe the switch will open before the car clears the top of the ramp. If we don't construct everything *exactly* right, the switch may open before the car goes past the transition from the sloping ramp to the top of the hill." (She drew the sketch shown in Figure 3.4.)

*Juan*: "Yes, early switch opening might be a problem. But, we can try my idea and just see if it works."

*Tina*: "Before we do that, though, how about this idea: We can connect a long screw thread to the drive shaft of the car and put a sliding nut of some kind (one that doesn't turn) along the shaft. As the shaft turns, it will thread through the nut and move it toward a normally-closed switch on the car body, opening it just after the right amount of distance is traveled. It might look something like this:" (She drew the sketch shown in Figure 3.5 and showed it to Juan.)

*Juan*: "Yeah! We could let it self-calibrate before each run by placing the car on the top of the hill with the nut just forcing the switch into the 'off' position, then manually roll the car backwards down to the starting point. As the nut travels back along the threaded rod, it will be placed in just the right starting position so that it trips the switch to off when the car gets to the top of the ramp." *He thought a moment more.* "How about a butterfly wing nut on the rotating shaft? We could buy one at the hardware store instead of having to make our own. One wing of the nut could ride against the car frame and prevent the nut from turning as the shaft turns. That way, the nut would be screwed down along the threaded shaft. The other wing of the nut could be used to press the switch." (Juan drew the sketch shown in Figure 3.6.)

*Tina*: "Yes. We can use one of those switches that has three terminals: normally open (NO), normally closed (NC), and common (COM)." Tina was describing a switch of the type shown in Figure 3.7. "The switch stays in the closed position, with COM connected to NC, and with NO connected to nothing, until the switch is pressed. When that happens, COM becomes connected to the NO terminal and NC is left unconnected."

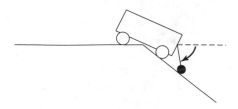

**Figure 3.4.** Premature opening of the switch.

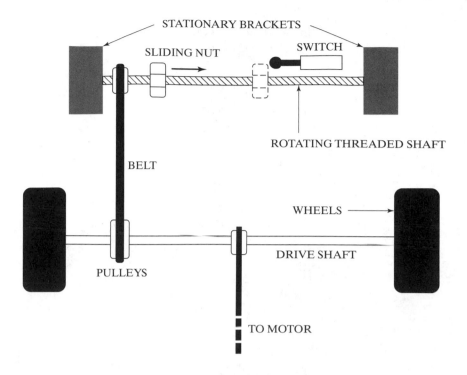

**Figure 3.5.** Nonturning nut moves along a rotating, threaded shaft.

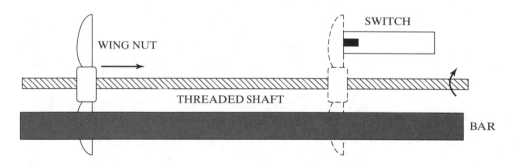

**Figure 3.6.** The wing nut concept.

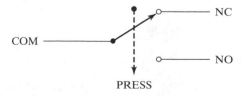

**Figure 3.7.** Switch showing normally open (NO) and normally closed (NC) contacts. The normal state of the switch refers to the condition where no force is applied to its pushbutton or lever arm.

After some thought, Tina continued. "You know, we don't have to run a separate threaded rod from the drive shaft. We can make the drive shaft *be* a threaded rod, like this." (Tina drew the sketch shown in Figure 3.8 and presented it to Juan.)

"That way, we'll have a simpler design, less friction, and an easy way to attach the wheels to the drive shaft with locking nuts."

*Juan:* "I like this idea much better than the switch idea. It seems more foolproof."

*Tina:* "On the other hand, if we use the threaded-rod idea and the wheels slip at all, the wing nut will still thread itself down the turning shaft, but the car won't be moving, so it will stop before it reaches the top of the hill."

*Juan:* "We could go back to our separate shaft idea and grease the shaft very well to reduce friction."

*Tina:* "No, that still won't solve the problem of the slipping wheels."

*Juan:* "Well, then, I have another idea. We could use a threaded shaft that is separate from the drive shaft and have it be turned by two idler wheels, rather than by the motor. The idler wheels will run along the track and turn the wing nut shaft but will be much less likely to slip because they won't be driven by the motor and will experience a much smaller torque." (Juan drew the sketch shown in Figure 3.9.)

"The idler wheels won't be connected in any way to the drive shaft or driven wheels. Even if the drive wheels slip, the only way that the idler wheels can turn is if the car is moving. If the idlers themselves get stuck, they'll drag along the track and *not* move the nut by the correct amount. But if the mechanism is well lubricated, sticking idler wheels should not be a problem."

*Tina:* "OK. Do you think we should consider an electronic timer that keeps the motor turned on for a fixed amount of time? We could experiment with the car and find out the exact amount of time needed for it to get to the top."

*Juan:* "Or how about an altimeter that measures the height of the car off the floor?"

*Tina:* "Or maybe a string that hooks to the base of the ramp, unwinds as the car goes up, and pulls a switch to the off position when the car gets to the top?"

*Juan:* "That's not allowed by the rules. All parts of the car have to lie inside the 1-meter marks at the end of 15 seconds."

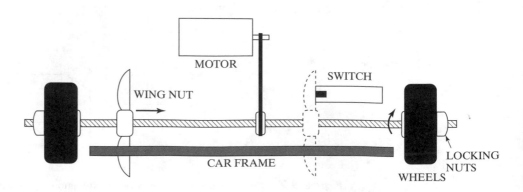

**Figure 3.8.** Threaded rod also serves the function of a drive shaft.

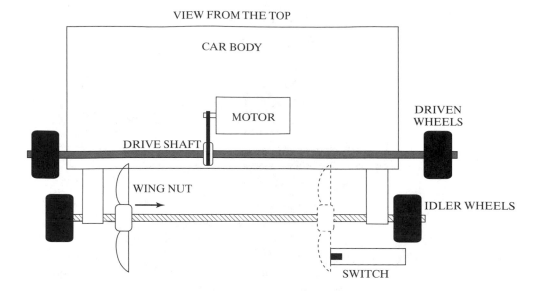

**Figure 3.9.** Idler wheels turn the threaded rod, which is not driven by the motor.

Tina checked the rules and confirmed that Juan was right. They both considered the altimeter idea and decided that it was not feasible. *If* they could find one at all, an altimeter with a resolution in the 1 meter range would be an expensive (and heavy) instrument indeed. Neither student wanted to design an altimeter from scratch. Their primary desire was to focus on the design of the vehicle itself. They also decided against the electronic timer idea for now. They were afraid that the car might travel at increasingly slower speeds from run to run as the battery ran down, and they remembered from their electronics class that the speed of the typical electronic timer is independent of its power supply voltage.

With their preliminary design concepts prioritized, they focused their attention on the two mechanical solutions: the trip switch and the concept of the sliding nut. They decided to build a few prototypes and try out the two basic ideas.

The foregoing conversation between Tina and Juan illustrates the principles of informal brainstorming. As each person stated a new idea, the other amplified upon it and came up with new ones, without prejudging each other. In the end, the two students condensed their list of ideas into one or two concepts that seemed feasible and they agreed to try them out in test experiments. The flow of conversation between Tina and Juan was appropriate for two people and even would have worked for three or four. Had their design team been larger, a formal brainstorming method, such as the idea trigger method discussed in the previous section, would have been more appropriate.

## PROFESSIONAL SUCCESS: WHAT TO DO WHEN ONE INDIVIDUAL DOMINATES A BRAINSTORMING SESSION

Suppose that you are the leader of a brainstorming session, and one member of your team dominates the conversation. That person may criticize participants, dismiss unconventional ideas, cut off speakers, or otherwise break the rules. When this situation occurs, it's your responsibility to keep the offender in line. Say to the group, "Hey, we need to stick to the formal rules of brainstorming. Let's institute a don't-speak-until-called-upon rule." This approach will tactfully short-circuit the behavior of the dominant person and maintain harmony among team members.

## PRACTICE!

1.  Conduct a one-person mini-brainstorming session and add as many ideas as you can to the final list of ideas compiled by Tina, Juan, Paul, and Karin during their idea trigger session. Allow yourself 4 minutes of brainstorm time to compile your ideas.

2.  A nonengineering friend complains about a pair of eyeglasses that keeps falling off. Give yourself 5 minutes of brainstorming time, and compile a list of as many ideas as you can for solving your friend's problem. After the 5 minutes are over, take a short break, then sort out your list, categorizing each idea with a "feasibility" rating of 1 to 5, with 5 the most feasible.

3.  Over a time span of 2 minutes, write down as many ways as you can for safely confining a dog to your backyard.

4.  Write down as many ideas as you can for designing a hands-free water faucet.

5.  As a way to save energy, you'd like to devise a method for reminding people to turn off lights. In a time span of 2 minutes, devise as many methods as you can to achieve this objective.

6.  Can brainstorming be used to solve math problems? Why or why not?

7.  Develop as many concepts as you can for a system that will automatically feed pellets to a pet hamster. Allow yourself 3 minutes of brainstorming time. If possible, work with a group of up to four people.

## 3.3   DOCUMENTATION: THE KEY TO PROJECT SUCCESS

Engineering design is never performed in isolation. Even the simplest of projects involves a designer and those who will benefit from the finished product. More often, a design effort involves considerably more individuals who worry about many facets of the product. The design of an automobile, for example, encompasses the work of mechanical, electrical, industrial, biomedical, and safety engineers. A consumer products company involved in the production of, say, cellular telephones, will bring together computer, electrical, mechanical, and manufacturing engineers in a multidisciplinary team that may include people from sales and marketing. Complex engineering projects are successful only if everyone on the design team communicates with everyone else at all phases of the design effort.

One way in which engineers communicate with each other is through careful record keeping. Good documentation is essential when engineers work in teams. When you work as a member of a design team, it's your responsibility to maintain a comprehensive collection of design concepts, sketches, detailed drawings, test results, redesigns, reports, and schematics. This *documentation trail* serves as a tool for passing information on to team members who may need to repeat or verify your work, manufacture your product from a prototype, apply for patents based on your inventions, or even take over your job should you be promoted or move to another company. Written records are also a good way to communicate with yourself. Many an engineer has been unable to reproduce design accomplishments or confirm test results due to sloppy record keeping. Indeed, one of the marks of a professional engineer is the discipline needed to keep accurate, neat, up-to-date records. Documentation should never be performed as an afterthought. If a project is dropped by one team member, the state of documentation should *always* be such that another team member can resume the project without delay. As a student of engineering, you should learn the art of record keeping and develop good documentation habits early in your career. Most companies, laboratories, and other technical institutions require their employees to keep records that document the results of their engineering efforts.

### 3.3.1 Paper Versus Electronic Documentation

Today, just about every piece of engineering documentation, with the exception of the engineer's notebook described in the next section, is produced on a computer. Examples of documents destined for preservation include word processed text, spreadsheets, schematics, drawings, design layouts, and simulated test results. Some engineers prefer to store information on disk so that it can be viewed on screen and printed out only as needed. Others prefer the older method of preserving documentation by printing everything on paper and storing the documents in a physical file cabinet. Whichever method you choose, you should follow the following important guidelines:

- *Organize your information:* It's important to store documentation in an organized and logical manner. If the project is small, its documentation should be stored in a single folder (paper or electronic). Larger projects may require a group of folders, each relating to different aspects of the project. The folders should be labeled and dated with informative titles such as "Propulsion System for XYZ Project" and kept in a place that will be easy to find should another team member need to use it.

- *Back up your information:* It's equally important to store a duplicate copy of all documentation. This guideline applies to written as well as electronic information. Fire, flood, theft, misplacement, and the all-too-common disk crash can lead to the loss of a project's documentation trail. Archival storage of records in a different physical location will help to keep a project on track should one of these catastrophes occur.

### 3.3.2 The Engineers' Logbook

One important vehicle for record keeping is the *engineer's logbook*, sometimes called the *engineer's notebook*. A well-maintained logbook serves as a permanent record that includes all ideas, calculations, innovations, and test results that emerge from the design process. When engineers work in a team, each team member keeps a logbook. When the project is brought to completion, all logbooks of all team members are placed in an archive and remain the property of the company. An engineering notebook thus serves as an archival record of new ideas and engineering research achievements *whether or not they lead to commercial use.* A complete logbook serves as evidence of inventorship and establishes the date of conception and "reduction to practice" of a new idea. It shows that the inventor (you!) has used diligence in advancing the invention to completion. In this respect, the engineer's logbook is more than just a simple lab notebook. It serves as a valuable document that has legal implications. When you work as an engineer, you have a professional responsibility to your employer, your colleagues, and to the integrity of your job to keep a good logbook.

The notebook shown in Figure 3.10 is typical of many used in industry, government labs, and research institutions. It has permanently bound and numbered pages, a cardboard cover, and quadrille lines that form a coarse grid pattern. A label fixed to the front cover uniquely identifies the notebook and its contents. The company, laboratory, or project name is printed at the top, and the notebook is assigned a unique number by the user. In some companies and large research labs, a central office assigns notebook numbers to its employees when the notebook is signed out.

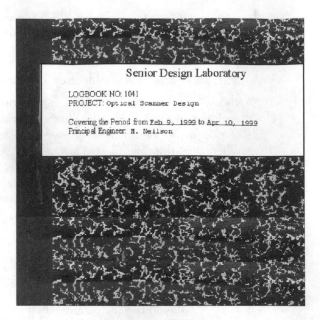

**Figure 3.10.** Cover label of a typical engineer's logbook.

The techniques for logbook use differ from those used in some science and introductory engineering classes where instructors encourage students to write things down first on loose scratch paper and then recopy relevant items into a neat notebook. This procedure is bad practice for a design engineer. Although notebooks prepared in this way are easier for instructors to grade, the finished notebook seldom resembles a running record of what actually occurred in the laboratory and is not especially useful for engineering design projects. Design is as much a *process* as it is a final product, and the act of writing down ideas as they emerge and of recording (and commenting on) events as they happen helps engineers to think and be creative. Also, keeping a record of what did *not* work is just as important as recording what did, so that mistakes will not be repeated in the future.

### 3.3.3 Logbook Format

An engineering logbook should be used as a design tool. Enter everything into your logbook, no matter how seemingly irrelevant. Write down ideas as you think of them, even if you have no immediate plans to pursue them. Keep an ongoing record of successes and failures. Record the results of every test—mechanical, structural, electrical, system, flight, or performance—even if the results may not be used in the final design. Stop to write things down. This habit will require discipline but will always be worth the trouble. Important information, including some you might otherwise have forgotten, will be in your logbook and will be at your fingertips when you need it.

Any logbook format that meets your needs and those of your team is suitable, as long as it forms a permanent record of your contributions to the design process. Ban loose paper from the laboratory. It is easily lost, misfiled, or spilled upon. Resist the temptation to reach for loose paper when you need to do a calculation. Instead of grabbing that pad to record information, draw a sketch, or discuss an idea, take the time to open your logbook. You'll be glad you did when those numbers and sketches you need are readily available. Unbound paper used for anything other than doodling has no place in an engineering laboratory.

### 3.3.4 Using Your Engineer's Logbook

As chief author of your logbook, you have the freedom to set your own objectives for its use. The following guidelines, however, are typical of those used by many engineers and design teams:

1. Each person working on a project should keep a separate logbook specifically for that project. All relevant data should be entered. When the logbook is full, it should be stored in a safe place specifically designated for logbook storage. In that way, everyone will know where to find the logbook when it's needed.

2. All ideas, calculations, experiments, tests, mechanical sketches, flowcharts, circuit diagrams, etc, related to the project should be entered into the logbook. Entries should be dated and written in ink. Pencil has a nasty habit of smudging when pages rub against each other. Relevant computer-generated plots, graphics, schematics, or photos printed on loose paper should be pasted or taped onto bound logbook pages. This procedure will help prevent loss of important data.

3. Logbook entries should outline the problem addressed, tests performed, calculations made, and so forth, but subjective conclusions about the success of the tests (e.g., "I believe...") should be avoided. The facts should speak for themselves. Logbook entries should not be a tape recording of your opinions.

4. The voice of the logbook should speak to a third-party reader. Assume that your logbook will be read by teammates, your boss, or perhaps someone from marketing.

5. In settings where intellectual property is at stake, the concluding page of each session should be dated and, where appropriate, signed. This practice eliminates all ambiguity with regard to dates of invention and disclosure. Important entries that signify key events in the design process should be periodically and routinely witnessed by at least one other person, and preferably two. Witnesses should endorse and date the relevant pages with the words, "witnessed and understood."

6. Logbook pages should not be left blank. If a portion of a page must be left blank, a vertical or slanted line should be drawn through it. Pages should be numbered consecutively and not be torn out. These measures are necessary should your logbook ever become part of a legal proceeding where the integrity of the information comes into question. Do not make changes using correction fluid. Cross out instead, This precaution will prevent you from creating obscure or questionable entries should your logbook be entered as legal evidence in patent or liability actions. Although this precaution probably won't be relevant to logbooks you keep for college design courses, it's a good idea to begin now so that the procedure becomes a career habit.

---

**EXAMPLE 3.3: AN ENGINEER'S LOGBOOK**

The following example illustrates proper use of an engineering logbook. Imagine that the logbook pages shown describe a self-propelled vehicle that you are designing for a student design competition called "Peak Performance." The first page, Figure 3.11, shows your preliminary sketch of a basic concept based on a vehicle that has the shape of a moving wedge. The second page, Figure 3.12, contains some calculations that estimate the battery drain as the vehicle moves up the contest ramp. The entries on the third page,

Figure 3.13, show a list of parts and materials to be purchased at the hardware store. These parts will enable you to build a prototype and test your vehicle's ability to climb the ramp.

DESIGN CONCEPT FOR DEFENSIVE STRATEGY.

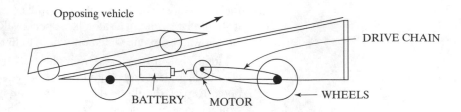

**Figure 3.11.** Logbook entry: Moving wedge concept for competition vehicle.

BATTERY POWER REQUIREMENTS

Estimate the weight of the vehicle:
$$0.9 \text{ kg} \times 10 \text{ N/kg} = 9 \text{ Newtons}$$

Compute the stored energy as the vehicle arrives at the top of the ramp:
$$9 \text{ N} \times 3 \text{ ft} \times 12 \text{ in/ft} \times 0.25 \text{ m/in} = 8 \text{ J}$$

Estimate the mechanical power. Assume vehicle takes about 7 sec to travel up the ramp:
$$8 \text{ J/7 sec} = 1.1 \text{ Watts}$$

Estimate the current drain on a 9-V battery (assume $P_{elec} = P_{mech}$ -- neglect losses for now):
$$I = P/V = 1.1 \text{ W/9V} = 0.12 \text{ A}$$

**Figure 3.12.** Logbook entry: Power consumption calculations.

PEAK PERFORMANCE DESIGN COMPETITION      2/7/99

PARTS LIST (to be purchased at hardware store)

Nuts and bolts (#8 × 1/2" with washers)
Wood screws (#6 × 3/4" long)
L-brackets (4):
Super Glue
Electrical tape
Solder (small cheap soldering iron? Or borrow?)
Switch (may have to go to electronic parts store)
Long threaded rod #10 thread size (will they have one?)
#10 thread wing nut

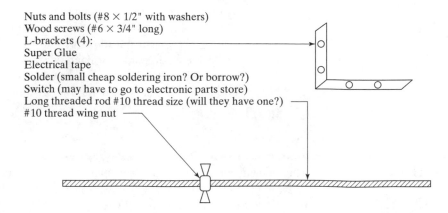

**Figure 3.13.** Logbook entry: List of parts to be purchased at the hardware store.

### 3.3.5   Technical Reports and Memoranda

Logbooks provide but one method by which a team keeps a good documentation trail. Engineers also communicate by writing technical reports at the significant milestones of a design project. A technical report describes a particular accomplishment and perhaps provides some project history or background material before explaining the details of what was achieved. The report may contain theory, data, test results, calculations, design parameters, or fabrication dimensions. Technical reports form the backbone of a company's or laboratory's technical database. Reports are typically stored in archival format, each with its own title and catalog number. Information for technical reports is easily gathered from logbooks that are accurate and up to date. When the time comes to write a journal paper, patent application, or product application note, the technical report becomes an indispensable reference tool. It is wise for engineering students to study techniques for writing technical reports in a clear and concise manner.

A technical report is also an appropriate way to explain why a particular idea did not work or was not attempted. Taking the time to write a technical report about a negative result or design failure can save considerable time should a design concept be revisited by engineers who were not present when the original project was undertaken.

### 3.3.6   Schematics and Drawings

Documentation does not always appear in the form of text. Graphical records, such as drawings, circuit schematics, photographs, and plots, also become part of the documentation trail. These items typically are created with the help of computer software tools. If the documents are to be stored electronically, then all files related to a particular project should be organized in a logical hierarchy. Some engineers choose to keep all files for a project in a single file folder on the computer. Others prefer to sort files by the applications that produce them (e.g., CAD drawings in one folder, spreadsheet files in another). Other engineers like to transfer all the computer files related to a given project to a single removable disk that can be stored in a physical file cabinet folder. If paper is chosen as the storage medium, then graphical output should be printed on paper and kept in a folder along with other written records. Regardless of which storage method is chosen, the information related to a particular project should be carefully preserved in a format that will prevent loss.

### 3.3.7   Software Documentation and the Role of the Engineering Notebook

Of all design endeavors, the writing of software is the one most prone to poor documentation. The revision loop of a software design cycle can be extremely rapid, because the typical software development tool enables a programmer to make small changes and test their effects immediately. This rapid-fire method of development invites poor documentation habits. Seldom does the software engineer find a good time to stop and document the flow of a program, because most pauses are short and change is frequent. As a result, the documentation for many software programs is added after the fact, if at all.

If you find yourself writing software, get into the habit of including documentation in your program as you go along. All software development tools provide a means for adding comment lines right inside the program code. Add them frequently to explain why you've taken a certain approach or written a particular section of program code. Explain the meaning of object names and program variables. Your in-program documentation should enable other engineers on your team to completely understand and take over the writing of your sections of the program simply by reading the comment lines. Good in-program documentation also will be invaluable to you should you need to modify your program at a later time. It's amazing how quickly a programmer can forget the internal logic of a program after setting it aside for only a short time.

If your program is destined for commercial sale, then good internal documentation and truly helpful "help" files are essential. Documentation included inside the program on a regular basis will easily translate into help files and an instruction manual when the need arises. One trick used by top-notch software developers is to write the help and instruction files as the program code is developed, rather than as an afterthought. Changes to the instruction manual can be made at the same time that changes are made in the program code. The abundance of commercial software packages with pathetic or poorly written help files or instruction manuals is testimony to generations of software engineers who have perpetuated a tendency toward poor documentation habits. If you master the skill of documenting software, your software products will be better utilized and more successful than those with poor documentation.

Although the keeping of engineering logbooks is less relevant to software development than to other types of engineering, logbooks still can play a special role. On the pages of your notebook, you can outline the overall flow of the program and the interconnections between its various modules. You can try out sketches for graphical user interfaces without first having to write actual computer code. You can draw plots of relational databases and make lists of the variables to be used in the software.

---

**PROFESSIONAL SUCCESS:   HOW TO KEEP GOOD RECORDS ALL THE TIME**

If you want to keep a good documentation trail, get into the habit of carrying your logbook with you wherever you go. In that way, it will be available whenever you have a thought or idea that needs recording. Buy a medium-size notebook that can fit easily into your backpack. Clip a pen right inside the front cover. Be sure to write your name and contact information on the front cover in case you misplace your logbook! A tiny, 3″ × 5″ bound notebook will do nicely. Although writing space will be limited because of the smaller size, you'll be more likely to carry it if it's not overly large.

---

## PRACTICE!

1.  Refer to the logbook calculations of Figure 3.12. Revise the estimate of the current drain on the battery if the vehicle weight is 2.1 kg.
2.  Refer to the logbook calculations of Figure 3.12. Convert all quantities to metric units and rewrite the logbook page.
3.  Refer to the logbook calculations of Figure 3.12. Revise the estimate of the current drain on the battery if the ramp height is 1.4 m and its length is 3.3 m.

### 3.3.8   The Importance of Logbooks: A Case Study

This case involved a small, Massachusetts-based biomedical engineering company called Abiomed (pronounced ab-ee-oh-med). Among its other heart-related projects, Abiomed is a leading contender for the development of an artificial human heart that will become a permanent replacement for individuals that could otherwise stay alive only with a human heart transplant. As noted on the company Website (*www.abiomed.com*), heart disease is the leading cause of death in the United States. Hundreds of thousands of people in the United States alone could be saved each year if transplants were available, yet only about 2,000 transplantable hearts become available in any given year. Hence, the need for a permanent artificial heart is real and widespread. Doctors at Jewish Hospital in Louisville, Kentucky installed the first Abiocor artificial heart in a human patient in July 2001.

Abiomed began its total artificial heart project in earnest in the early 1990s. The major components of the heart system, depicted in Figure 3.14, include the central pump, interconnecting tubes that connect to the subject's principle blood vessels, and a system for transferring power through the skin from a battery pack worn outside the body. One key feature of the power transfer process, the lack of wires piercing the skin, is essential for the long-term efficacy of the artificial heart, because skin perforations are prime entry points for infection and require constant medical supervision in a skilled-care facility. The Abiomed artificial heart is being designed to enable the patient some semblance of a normal, mobile, home life. The system for transferring electrical power, called the transcutaneous energy transfer device, or "TET," is essentially a pair of concentric, high-frequency magnetic coils—one implanted under the skin, and one worn outside the skin. (Transcutaneous means "through the skin.") Abiomed wished to have its engineers fully focus on the daunting task of developing the heart pump itself, and so it hired another biomedical engineering company, World Heart Corporation of Ottawa, Canada, to design the energy transfer module.

After about 4 years of effort, World Heart was still unable to produce a TET that met Abiomed's stringent technical specifications. Although World Heart claimed to be converging on a solution, Abiomed was not convinced that a satisfactory TET device was imminent. Faced with an impending critical animal test that would determine future funding of its entire heart project from the National Institutes of Health, Abiomed decided to sever its reliance on the World Heart TET. The Abiomed CEO instructed one of his engineers, Dr. Z, to develop a home-grown TET device as quickly as possible. Dr. Z is a very capable fellow, and after only 4 months of effort, he succeeded in designing and testing a fully-working version of a TET device that met all the requirements of Abiomed's impending heart-pump test.

This development led to a lawsuit by World Heart, who claimed that Abiomed could not have developed its own TET device in the mere time span of 4 months without having stolen secrets and technology from World Heart. After all, World Heart claimed, its engineers had worked on the project for 4 years and were only just beginning to converge on a successful solution. How could the Abiomed engineer have designed a superior TET in only 4 months? Abiomed countered with a claim that its short path to success was due solely to the high competency of its engineer, and that, in fact, it had taken special precautions to ensure that nothing would be stolen from the previous World Heart design effort. The suit went to court and, in the end,

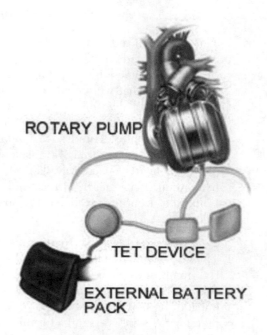

ROTARY PUMP

TET DEVICE

EXTERNAL BATTERY
PACK

**Figure 3.14.** Total artificial heart system. (*Graphic courtesy of Abiomed, Inc.*)

Abiomed prevailed. The jury recognized that Abiomed's TET design, while performing the same basic transcutaneous energy-transfer function as the World Heart device, was completely different with regard to all details of implementation. The Abiomed device, the jury concluded, used different circuits, materials, magnetic construction, and semiconductor components.

During the trial, a key component of Abiomed's defense was the logbook kept by the engineer who designed the TET. This logbook was used to prove that Abiomed had designed its own, independent version of the TET. Following the company's customary logbook policy, as well as sound engineering practice, Dr. Z had kept careful records of his TET design, having entered every design concept, schematic, circuit layout, and test result that emerged during the design process. He had even noted the various circuit configurations that had blown up on the test bench before his first working model emerged. Each page of his logbook had been dated, and each lab session involving successful tests had been signed and countersigned by another Abiomed engineer. During trial, entire pages of Dr. Z's logbook were reproduced and projected on a large screen for the jury to see. Dr. Z's logbook played a crucial role in the success of Abiomed's legal defense.

The presentation of Dr. Z's logbook in court, while critical to the outcome of the trial, did not proceed flawlessly, however. Several pages of the logbook involving work done in mid January 1996, had been incorrectly dated with the year 1995. Following a common mistake that many individuals make when the year changes, Dr. Z had, without thinking, hastily written the year from the previous month of December. The lawyers for World Heart were quick to seize upon this error as prima fascia evidence that something was "fishy." They claimed Dr. Z had forged portions of his logbook in an attempt to present a false picture of his accomplishments. In the end, the jury was not convinced and realized that Dr. Z's error was nothing more than a common calendar mistake. Nevertheless, this seemingly small lack of attention to detail in logbook procedure put the case against Abiomed in jeopardy for a time.

---

**PROFESSIONAL SUCCESS: DEVELOPING GOOD LOGBOOK HABITS**

Most of us follow routines without thinking. When we wake up in the morning, we brush our teeth. When we eat, we instinctively reach for a clean plate. When we get into a car, we (hopefully) buckle our seatbelts automatically. An as engineer, the urge to write things down in your logbook should become as instinctive as these other common tasks. In contrast to these personal health procedures, however, we engineers are not trained from childhood to record our experiences in a notebook. Developing this instinct requires practice, but it can become part of your routine over time. When personal computers and the Internet first came into being, most people did not think very much about the novelty of e-mail. Now checking one's mail has become a daily routine for most. You developed this unnatural skill by practicing it over time. It should be the same with your engineer's logbook. Force yourself to get into the habit of using your logbook whenever you practice design. Over time, it will become as natural a gesture as brushing your teeth.

---

## 3.4 PROJECT MANAGEMENT: KEEPING THE TEAM ON TRACK

Even the simplest of design project must be properly managed if it is to be successful. A systematic approach to design is always preferable to a random, hit-or-miss approach. While the subject of project management could (and does) occupy the contents of entire books and the curricula of programs in business administration, three project management tools—the *organizational chart*, the *time line,* and the *Gantt chart*—form a basic set that should be understood by all engineers.

### 3.4.1 Organizational Chart

When engineers work on a team-oriented design project, some hierarchy among individuals is necessary. It would be nice if an engineering team could always function as a simple group of colleagues, but inevitably some team members will be burdened more than others, and some tasks will fall between the cracks, unless everyone's responsibilities are clearly spelled out. One vehicle for specifying the management structure of a design team is the *organizational chart*. An organizational chart indicates who is responsible for each aspect of a design project. It also describes the hierarchy and reporting structure of the team. Figure 3.15 illustrates a simple organizational chart that might be used by students in a vehicle design competition. In this particular case, Tina acts as the team leader, but she, in turn, reports to the course teaching assistant for leadership and guidance. In the corporate world, where the structuring of workers and bosses can become complex, organizational charts are essential because each employee must understand to whom he or she reports and must know the responsibility chain from upper management on downward. Note that organizational charts are used by all sorts of companies, not just engineering firms. Figure 3.16, for example, outlines the organization of the transportation department of a large American city. The features of this chart are identical to those found in the charts of large engineering companies.

### 3.4.2 Time Line

Time management is critical to the success of any design project. In a perfect world, engineers would have as much time as needed to work on all aspects of a design project, but in the real world, deadlines have a nasty habit of creating pressure to "get the product out the door." Demands for demonstrations of progress, prototype tests, something for "sales and marketing to show," and the pressures of corporate life require that an engineer develop a sense of how much time will be needed for each aspect of product development. A *time line* is a valuable tool for an engineer who wants to keep a project on schedule. A time line is simply a linear plot on which each of the various phases of a design project is assigned a milestone date. If a given task is in danger of not being completed before its designated milestone date, it is the job of the engineer to allocate more time, and overtime if necessary, to the task so that it can be completed on schedule. A typical time line, in this case one prepared by students designing a vehicle for a class design competition, is shown in Figure 3.17.

### 3.4.3 Gantt Chart

When a project becomes complex and involves many people, a simple time line may be inadequate for managing all aspects of the project. Similarly, if the project's various parts are interdependent, so that the completion of one phase depends on the success of others, the *Gantt Chart* of Figure 3.18 is a more appropriate time-management

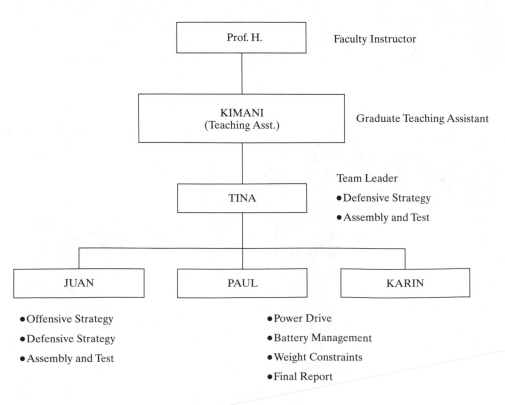

***Figure 3.15.*** Organizational chart showing responsibilities of a Peak Performance vehicle design team.

tool. The Gantt chart is simply a two-dimensional plot in which the horizontal axis is time measured in blocks of days, weeks, or months, and the vertical axis represents either the tasks to be completed or the individuals responsible for those tasks. Unlike the simple one-dimensional time line, which displays only the milestone dates for each phase of the project, the Gantt chart shows how much time is allotted to each task. It also shows the time overlap periods that are indicative of the interdependency-between the various aspects of the project. When a particular task has been completed, it can be shaded in on the Gantt chart, so that the status of the project can be determined at a glance.

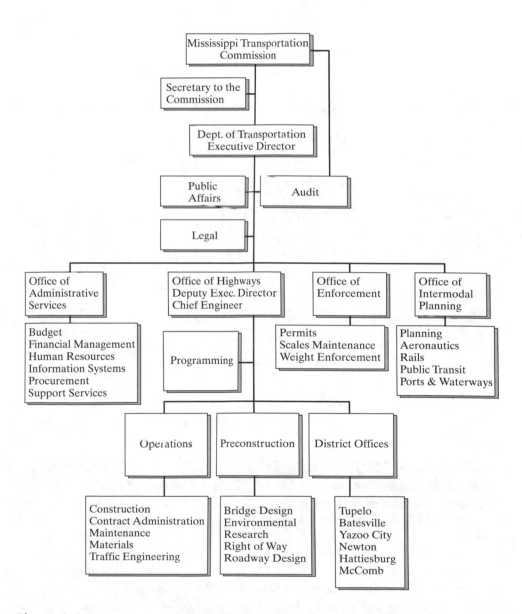

**Figure 3.16.**  Organizational chart of the Mississippi Department of Transportation. (*Chart courtesy of MDOT.*)

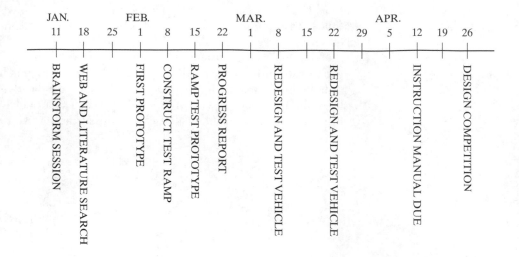

**Figure 3.17.** Time line for scheduling tasks for the Peak Performance vehicle design team.

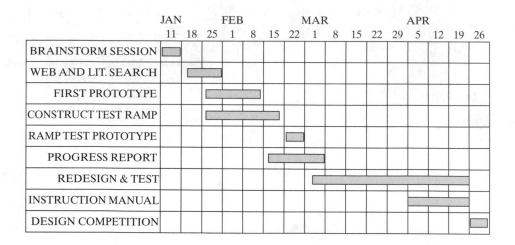

**Figure 3.18.** Gantt chart provides a more comprehensive, two-dimensional method of scheduling the tasks for the Peak Performance vehicle design team.

---

**PROFESSIONAL SUCCESS: THE LAWS OF TIME ESTIMATION**

How long will it take to perform a particular design task? How much time should I allot to each segment of the design cycle? The following three Laws of Time Estimation will help you to determine the time required for elements of the design process:

1. Everything takes longer than expected.

2. If you've worked on something similar before, estimate the amount of time required to finish the task. The actual amount of time required will be about four times longer.

3. If you've *never* worked on something similar before, estimate the amount of time required to finish the task. The actual amount of time required will be equal to the next highest time unit: Something estimated to take an hour will really take a day; something estimated to take a day will take a week, etc.

**PRACTICE!**

1. Devise an organizational chart and time line for building a float for the homecoming parade at your college or university.

2. Develop a time line for completing your course requirements over the time span of 4 academic years.

3. Develop a Gantt chart for hosting an educational conference on engineering design. Consider all needed arrangements, including food, transportation, lodging, and meeting facilities.

4. A large nonprofit organization is planning a walkathon in which about 3,000 people will walk a total of 60 miles in 3 days and will sleep in tent camps each night. Develop an organizational chart for all persons involved in planning and implementing this event.

5. A student organization is planning to enter an international solar car competition (see, for example, *www.winstonsolar.org*). Develop a Gantt chart that will guide the team in the design and construction of a competition vehicle.

**KEY TERMS**

| | | |
|---|---|---|
| Teamwork | Brainstorming | Documentation |
| Project Management | Time Line | Organizational Chart |
| Gantt Chart | | |

**PROBLEMS**

*Brainstorming:* Use brainstorming methods to generate solutions to the following problems:

1. You are given a barometer, a stop watch, and a tape measure. In how many different ways can you determine the height of the World Trade Center in New York City?

2. Design a sensing mechanism that can measure the speed of a bicycle.

3. Many international airline flights still allow smoking in the rear seats of the aircraft. Design a system that will remove or deflect smoke from the front seats of the aircraft.

4. You are given an egg, some tape, and several drinking straws. Using only these materials, design a system that will prevent the egg from being broken when dropped from a height of 6 ft (2 m).

5. Devise as many different methods as you can for using your desktop computer to tell time.

6. Design a system for washing the inside surfaces of the windows of a large aquarium (the kind the public visits to see large sea creatures) from the outside.

7. Design a system to be used by a quadriplegic to turn the pages of a book.

8. Devise a system for automatically raising and lowering the flag at dawn and dusk each day.

9. Design a system that will automatically turn on a car's windshield wipers when needed.

10. Develop a device that can alert a blind person to the fact that water in a pot has boiled.

11. Devise a system for lining up screws on an assembly-line conveyor belt so that they are all pointing in the same direction.

12. Develop a method for detecting leaks in latex surgical gloves during the manufacturing process.

13. Devise a method for deriving an electrical signal from a magnetic compass so that it can be interfaced with a computer running navigational software.

14. Given a coil of rope and eight poles, devise a method to build a temporary emergency shelter in the wilderness.

15. Devise an alarm system to prevent an office thief from stealing the memory chips from inside a personal computer.

16. Imagine custodial workers who are in the habit of yanking on the electric cords of vacuum cleaners to unplug them from the wall. Devise a system or device to prevent damage to the plugs on the ends of the cords.

17. Develop a system for automatically dispensing medication to an elderly person who has difficulty keeping track of schedules.

18. Develop a system for reminding a business executive about meetings and appointments and changes in schedule that originate from the home office. The executive is always on the go, but can carry a variety of portable devices and gadgets. Feel free to use your knowledge of existing communications systems and technology, if necessary.

19. Devise a system that will agitate and circulate the water in an outdoor swimming pool so that a chlorine additive will be evenly distributed. Assume that an electrical outlet is available at the site of the pool.

20. Devise a system that will allow a truck driver to check tire air pressure without getting out of the vehicle.

***Documentation:***

21. Begin to keep a logbook of your class activities. Enter sketches and records of design assignments, inventions, and ideas.

22. Pretend that you are Alexander Graham Bell, the inventor of the telephone. Prepare several logbook pages that describe your invention.

23. Pretend that you are Marie Curie, the discoverer of the radioactive element radium. Prepare several logbook pages that describe the activities leading to your discovery.

24. Pretend that you are Dr. Zephram Cockrane, the inventor of plasma warp drive on the television and movie series *Star Trek*. Prepare several logbook pages that describe your invention. (*Star Trek, First Contact: www.startrek.com/library/movies/viii_main.html*)

25. Imagine that you are Elias Howe, the first inventor to perfect the sewing machine by putting the eye of the needle in its tip. This innovation made possible the bobbin system still in use in sewing machines today. Prepare several logbook pages that describe your invention and its initial tests.

26. Reconstruct logbook pages as they might have appeared for the person inventing the common paper clip.

27. Imagine that you are Dr. Z, the engineer involved in the Abiomed case described in Section 3.3. Sketch the logbook pages that outline the basic operating concept of your transcutaneous energy transfer device.

28. The invention of the incandescent light bulb is largely attributed to the famous American inventor, Thomas Edison. Reconstruct the logbook pages that Edison may have kept describing his classic design efforts.

29. The cotton gin was developed by an American inventor, Eli Whitney, around 1800. This invention had a profound effect on the economics and history of the early United States. Reconstruct logbook pages in which Whitney outlines the basic features and development of his invention. Note that patent law in America was in its infancy around the time that Whitney did his work on the cotton gin.

30. Samuel F.B. Morse, inventor of the telegraph in the 1830s and the pioneer who launched the world's first "information age," was actually an artist by profession when he developed his classic invention. During a long voyage home from study in France, he passed his time thinking about conversations he had heard concerning ongoing experiments in Europe on electricity and magnetism. He developed his ideas for the telegraph while returning to the United States. Sketch out the logbook pages that Morse may have kept during his long sea voyage.

31. The first calculator was designed by Jack Kilby, an engineer for Texas Instruments. Look up the history of this inventor on the Web, and see if you can reconstruct the probable appearance of one or more pages from his logbook.

32. Sketch out a logbook page that describes a design concept for a recumbent bicycle.

33. Prepare a logbook page that describes the inner workings of a common CD player.

### Project Management:

34. Suppose that you've been given the assignment to write a research paper on the history of human air flight. Develop a time line for completing this comprehensive research assignment.

35. Create a Gantt chart for your own hypothetical entry into a national solar-powered vehicle design competition.

36. Imagine that you work for a company that is designing an electric car for commercial sale. Create an organizational chart for the company and a Gantt chart for designing the vehicle's drive train.

37. Choose an engineering company with which you are familiar or in which you have an interest. Develop an organizational chart for the company. Information about a company's structure and personnel often can be found on the company's Web site.

38. Imagine that you wish to start your own company to write software tools for doing business on the Web. Create an organizational chart that outlines the positions you'll need to fill to get the company started.

39. Develop a time line for the completion of the prototype of an automobile powered from fuel cells rather than an internal combustion engine.

# 4

# Engineering Design Tools

If you were to hire a carpenter to install some new kitchen cabinets, you would not expect that person to arrive on the job empty-handed. As a professional tradesperson, the carpenter would bring along an array of power and hand tools, including saws, hammers, drills, and screwdrivers. The carpenter probably also would bring some common fasteners and raw materials as needed to finish the job. Likewise, you would not expect to visit a doctor's office without confronting an array of medical diagnostics, including stethoscope, tongue depressors, blood pressure cuff, reflex hammer, and examining table. Like these other professionals, engineers rely on numerous tools to aid in all facets of the design process. While some of an engineer's tools literally can be carried around in a toolkit — calculator, mechanical pencils, and laptop computer, for example — many fall into the category of *knowledge tools*. A knowledge tool is defined as a practice or methodology that the engineer has learned on the job or in school. Other tools exist in the form of software programs developed to help engineers address specific classes of problems. The purpose of this chapter is to highlight some of the more important knowledge and software tools that are found in an engineer's toolkit.

## SECTIONS

- 4.1   Estimation
- 4.2   Significant Figures, Dimensioning, and Tolerance
- 4.3   Prototyping and Breadboarding
- 4.4   Reverse Engineering
- 4.5   Computer Analyses
- 4.6   The Internet
- 4.7   Spreadsheets in Engineering Design
- 4.8   Solid Modeling and Computer-Aided Drafting
- 4.9   System Simulation
- 4.10   Electronic Circuit Simulation
- 4.11   Graphical Programming
- 4.12   Microprocessors: The "Other" Computer

## OBJECTIVES

*In this chapter, you will:*

- Examine the role of the computer in engineering design.
- Learn when and when not to use the computer.
- Discuss several examples of computer use for analysis, data collection, and real-time control.

## 4.1    ESTIMATION

Engineering design and estimation go hand in hand. When beginning any new design task, it's always a good idea to test for feasibility by doing rough calculations of important quantities and parameters. A paper-and-pencil or simple hand calculator analysis of a proposed strategy may eliminate gross inconsistencies before the actual construction process begins. The calculations need not be elaborate or precise. In the age of programmable calculators and computers, students sometimes feel that answers with lots of digits imply better or more accurate answers. In many cases, however, "back of the envelope" calculations done by hand (and recorded in your engineer's logbook) are all that are required to determine the soundness of a design strategy.

**EXAMPLE 4.1: ESTIMATING POWER FLOW FROM A BATTERY**

The following example illustrates the usefulness of estimation as a design tool. Suppose that you and your teammate have decided to build a car that is battery operated and propelled by an electric motor as part of a vehicle design competition called "Peak Performance." This contest requires two opposing vehicles to travel up the sides of a 1.5-m two-sided ramp to a height of 90 cm in an attempt to get as close as possible to the top. In an effort to conserve weight, you have decided to operate the vehicle from a single 9-V "transistor radio" battery, if possible. This design choice takes advantage of the fact that each run up the ramp, as specified in the contest rules, is short in duration (no more than 15 seconds), and teams are allowed to change batteries between runs. Your strategy will be to heavily tax each battery to its maximum output and change it between each run up the ramp. In order to determine whether such a strategy is feasible, you must estimate the power to be delivered by the battery as the vehicle travels up the ramp. If this required power exceeds the amount available from a single battery, you will need to alter your design strategy and use two or more batteries.

*Calculate the Power Required from the Battery*

You hope to limit your vehicle weight to 0.9 kg (about 2 pounds) which is less than half the maximum value of 2 kg allowed by the rules. A simple calculation will reveal the energy needed to lift a vehicle of this weight from the ground to the top of the ramp. The time duration over which this energy is extracted determines the power flow to the vehicle. Ultimately, this power must come from the battery in electrical form. The motor's job is to convert electrical power into mechanical power. Thus, the electrical power entering the motor will have to equal the mechanical power transmitted to the wheels plus any electrical or mechanical losses in the drive train. This power-flow relationship is summarized in Figure 4.1.

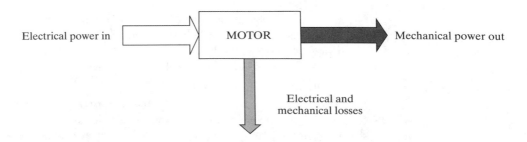

**Figure 4.1.**    Power-flow diagram.

*Determine the Weight of the Vehicle*

First, you compute the downward force $F_y$ on the vehicle, noting that the gravitational force (weight) will equal the mass of the vehicle times the gravitational constant:

$$F_y = -mg. \tag{4-1}$$

Here, $g$, the gravitational constant, has a magnitude of about 10 newtons per kg. The weight of a 0.9-kg vehicle thus will be (0.9 kg)(10 N/kg) = 9 newtons.

*Energy Equals Force Times Distance*

As the car is propelled to the top of the ramp, the mechanical energy supplied by the wheels will be equal to the net gain in the car's stored potential energy. This quantity will be equal to the magnitude of the gravitational force on the vehicle times the vertical distance traveled. The top of the ramp lies 90 cm above the floor; hence, the net gain in potential energy becomes

$$E = F_y \times y = (9 \text{ N})(0.9 \text{ m}) = 8 \text{ joules}. \tag{4-2}$$

The net gain in potential energy also can be expressed as the vector *dot product* of the total force **F** and the trajectory **s** of the vehicle as it travels up the ramp. The ramp is inclined at an angle $\theta$ relative to the vertical, where $\theta$ can be computed from the ramp specifications. As specified in the contest rules, the vertical rise of the ramp is 90 cm, and the path length traveled by the vehicle is 150 cm. Hence,

$$\cos\theta = \frac{\text{height}}{\text{path length}} = \frac{90 \text{ cm}}{150 \text{ cm}} = 0.6 \tag{4-3}$$

or

$$\theta = \cos^{-1} 0.6 = 53°. \tag{4-4}$$

The potential energy stored in the vehicle then becomes

$$E = \mathbf{F} \cdot \mathbf{s} = (9 \text{ N})(1.5 \text{ m})(\cos 53°) = 8 \text{ joules}.$$

This result is the same one computed in Equation 4-2.

*Compute the Mechanical Power*

The power flow is equal to the energy per unit time measured in joules per second. For estimation purposes, let's assume that the run up the ramp lasts about 7 seconds, or half the allotted time of 15 seconds. The mechanical power supplied by the wheels then can be estimated by dividing the stored energy by the time required to climb the ramp:

$$P_{\text{mech}} = \frac{E}{t} = \frac{8 \text{ J}}{7 \text{ s}} = 1.1 \text{ watts}. \tag{4-5}$$

A calculation such as this one should always be examined to make sure that the answer is reasonable. As a basis for comparison, consider a small plug-in night-light that draws about 4 watts. Expecting the car to draw about one quarter of that amount for 7 seconds indeed seems reasonable. The answer is believable.

*Compute the Required Electrical Power*

The electrical power supplied by the battery to the motor must *at least be* equal to the mechanical power required by the wheels plus any losses that may occur in the motor and drive train (e.g., the gears, belts, pulleys, bearings, and axle mounts). Neglecting these losses, you arrive at the simple conclusion that $P_{\text{mech}} = P_{\text{elec}}$. The electrical power supplied by the battery will be equal to its voltage times the current drawn out of the battery, i.e.,

$$P_{\text{elec}} = V\,I. \tag{4-6}$$

Hence, at a fixed voltage of 9 V and a power drain of 1.1 W, the battery will need to supply a current of

$$I = \frac{P}{V} = \frac{1.1\,\text{W}}{9\,\text{V}} = 120\,\text{mA}. \tag{4-7}$$

Your estimated calculations are complete. Next, you decide to obtain information about batteries to see whether or not a standard 9-V battery can supply a current of 120 mA. Accessing the Duracell™ and Eveready™ Web sites (*www.duracell.com* and *www.eveready.com*) you find that the typical 9-V battery can supply about 100 mA of current for short periods of time (about 1 hour or less.) This level of battery performance places your design specification at the border of feasibility. You can decide to stay with one battery only, possibly taxing it to its limit, or change your design specifications to include two batteries, each providing half the required power. For the moment, you decide that one battery is the better choice from the point of view of vehicle size and weight. The 100-mA battery capacity is a general guideline, not an absolute limit. If you stick with your strategy of replacing the battery after every run, it might perform adequately. Also, if you slow the vehicle to increase the run time to 10 seconds, the power required will be reduced to 0.8 W, and the current required will be reduced to 90 mA.

*On Second Thought . . .*

After reviewing your calculations and assumptions, you and your teammate realize that you have neglected all losses in the system. In reality, the conversion efficiency from electrical to mechanical power will be far from perfect. According to your professor, one might expect a power conversion efficiency of up to 90 percent from a well-designed, expensive motor, but you've decided to buy an inexpensive motor from a local electronics supplier to save money. Similarly, no more than about 60 percent of the converted mechanical power supplied by the motor shaft will show up at the wheels because of frictional losses in the gears, drive belts, and axle mounts, leaving only about 50 percent of the power taken from the battery to actually propel the vehicle. The two-battery approach seems to be the more reasonable choice after taking losses into account. This sort of re-evaluation is common in the engineering design process.

## PROFESSIONAL SUCCESS: BE WILLING TO MODIFY YOUR CONCLUSIONS WHEN NECESSARY

Despite the time and effort that go into making design decisions, a good engineer knows when it's time to admit an oversight and change a basic design decision. One example of this principle can be found in the standard automobile battery. Its voltage value was set at 12 V around 1940. Prior to that time, automotive engineers had agreed upon a battery voltage of 6 V, equal to the series combination of three lead-acid cells. The increasing electrical demands inside vehicles to power things such as headlights and radios led to increased power requirements. Keeping the standard voltage at 6 V would have meant large currents (and impractical thicker wires) running throughout the vehicle. Engineers from all major car makers agreed to increase the standard battery voltage to 12 V, thereby halving the needed current flows for the same amounts of electrical power.

Now, in the first decade of the 2000s, the automotive industry is undergoing another change in design strategy. The electrical loads on modern cars have increased dramatically over the past several years as consumers have demanded more and more accessories. Many of a modern car's safety and fuel-economy systems — antilock brakes, fuel injection, traction steering, and zoned air conditioning, for example — run on electrical power rather than on mechanical power from the gasoline engine. A move is underway to increase the standard battery voltage to 48 V so that, when compared to 12-V systems, current flow requirements will be reduced by a factor of four throughout the vehicle.

**EXAMPLE 4.2: ESTIMATING THE VOLUME OF PAINT NEEDED TO COAT A LARGE OBJECT**

This example, which comes from the real world of automobile manufacturing, illustrates the usefulness of estimation as an engineering design tool. Imagine that you work for a company that makes automobiles. The head of manufacturing thinks that the company could save a lot of money by abandoning conventional painting techniques and instead adopting a finishing process that impregnates color right into the body material during manufacturing using an electrostatic powder coating and baking technique. The comparison is a tough call. Labor and equipment depreciation, not materials, are the main costs involved in most manufacturing processes. The overhead at a typical fabrication facility, including benefits, insurance, physical plant (the cost of keeping the factory open so that workers can do their jobs), and depreciation can run anywhere from 60 percent to 200 percent of salaries and other direct costs. Adopting the labor-saving powder-coating method, therefore, seems like the better choice. On the other hand, in a large factory that can justify the initial equipment cost, cars and other large consumer items can be painted by robot, thereby also eliminating labor costs. A comparison between powder and paint coating methods thus reverts to a cost-of-materials comparison only.

The head of manufacturing has asked you, a manufacturing engineer, to estimate the total cost of conventional paint needed to cover the vehicle. How can you arrive at such an estimate? The steps required are outlined in the following discussion:

*Draw a Rough Sketch of the Surfaces to Be Painted*

As a first step, you should draw a rough sketch of the car body on paper. One such sketch that depicts the various car surfaces is shown in Figure 4.2. The largest areas to be painted include the hood, trunk, roof, and two side fenders. The paint required to cover the posts of the window frames is negligible.

*Estimate the Area of Each Section*

Next estimate the area of each separate section of the car. The hood forms an approximate 1.2 m × 1.2 m square for a total area of about 1.4 square meters. The trunk is also nearly rectangular, measuring about 1.2 × 1.5 m, for an additional 1.8 square meters. The doors are about 1 m long by 0.8 m tall, for a total area of 0.8 square meters each. The dimensions of the roof are about 1.4 m long by 1.2 m wide, for total of 1.7 square meters. For estimation purposes, each of the fenders can be modeled by one of the shapes shown in Figure 4.3. The surface area of the windows need not be counted, because they are not painted. The area of each fender can be calculated from the area formula for a trapezoid:

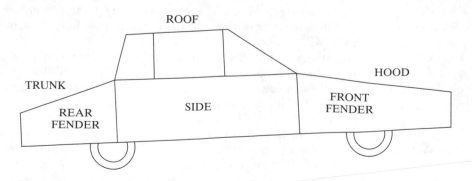

**Figure 4.2.**   Rough sketch of car body.

SIDE FENDER

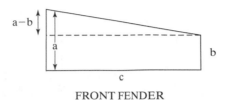

FRONT FENDER

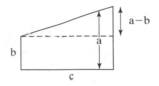

REAR FENDER

**Figure 4.3.** Estimated shape of side fenders.

$$A = bc + \frac{c(a-b)}{2}. \tag{4-8}$$

If you assume reasonable numbers for the dimensions of the fenders; for example: $a = 0.8$ m, $b = 0.4$ m, and $c = 1.5$ m for the front of the car, and $a = 0.8$ m, $b = 0.4$ m, and $c = 1$ m for the rear, then you can compute the individual area estimates and sum them to obtain an estimate for the total surface area of the car:

Hood: $1.2 \text{ m} \times 1.2 \text{ m} = 1.4 \text{ m}^2$ (4-9)

Trunk: $1.2 \text{ m} \times 1.5 \text{ m} = 1.8 \text{ m}^2$
Doors: $2 \times 1 \text{ m} \times 0.8 \text{ m} = 1.6 \text{ m}^2$
Roof: $1.4 \text{ m} \times 1.2 \text{ m} = 1.7 \text{ m}^2$
Front fenders: $2 \times (0.4 \text{ m})(1.5 \text{ m}) + (1.5 \text{ m})(0.8 \text{ m} - 0.4 \text{ m}) = 1.8 \text{ m}^2$
Rear fenders: $2 \times (0.4 \text{ m})(1 \text{ m}) + (1 \text{ m})(0.8 \text{ m} - 0.4 \text{ m}) = 1.2 \text{ m}^2$
**Total area**: $1.4 \text{ m}^2 + 1.8 \text{ m}^2 + 1.6 \text{ m}^2 + 1.7 \text{ m}^2 + 1.8 \text{ m}^2 + 1.2 \text{ m}^2 = 9.5 \text{ m}^2$

The result is equal to about 10 square meters. (Round numbers are appropriate because only a rough estimate is required.)

*Multiply by the Thickness of the Paint*

Next, you can estimate the volume of paint required to cover the car by multiplying its surface area by the paint thickness. The latter usually is measured in mils (1 mil = 0.001 inch = 0.025 mm = $25 \times 10^{-6}$ m). A very light coating of paint is typically about 0.5 to 1 mil thick, while a very heavy coating might be as thick as 7 or 8 mils. The choice of thickness for estimation purposes will depend on the paint application and whether you want your estimate to be on the high side or the low side of the actual value. Suppose that you choose an average value of 4 mils. For this thickness, the volume of paint required to coat the car can easily be calculated:

Volume = area × thickness = $(10 \text{ m}^2) \times (4 \text{ mils}) \times (25 \times 10^{-6} \text{ m/mil}) = 0.001 \text{ m}^3$, or about 1 liter.

## PRACTICE!

1. Determine the power required of the battery in Example 4.1 if the ramp is only two-thirds as high.

2. Determine the power required of the battery in Example 4.1 if a set of four 1.5-V AAA cells is used instead of a single 9-V battery.

3. How much mechanical power can be derived from a 6-V electric motor that is 85 percent efficient if its maximum allowed current is 1 A?

4. How much internal heat will be generated by an electric motor that is 60 percent efficient if it provides 20 W of mechanical power to an external load?

5. Estimate the amount of paint required to cover a single wooden pencil.

6. Estimate the physical length of a 90-minute audio cassette tape.

7. Estimate the number of platforms needed to build a scaffolding shell that encircles the Washington Monument.

8. Estimate the volume of water contained within the supply pipes of a one-story, single family house. How much energy is required to heat this water from 20°C (room temperature) to 80°C (shower temperature)? What is the significance of this energy quantity?

9. Estimate the number of staples in the loading clip of a standard office stapler.

10. You work for a company that paints houses in the summer. Your assignment is to paint a garage that measures 20 ft × 30 ft × 10 ft tall. How many gallons of paint (primer and finish coat) should you buy?

11. Estimate the volume of paint required to cover a Boeing 747 airplane.

12. Estimate the volume of ink required to print an 8.5 × 11 inch paper document in Ariel 10 point font, single-spaced type, with 1-inch margins. How many such pages do you think you could print from a typical black inkjet cartridge?

13. Estimate the energy required to heat the water needed to take your morning shower.

14. Estimate the volume of take-out coffee consumed by the national population each day, then work backwards to estimate the total weight of paper needed to make all those coffee cups.

## 4.2 SIGNIFICANT FIGURES, DIMENSIONING, AND TOLERANCE

The accuracy of any number used in estimations or other technical calculations is specified by the number of *significant figures* that it contains. A significant figure is any nonzero digit or any leading zero that does not serve to locate the decimal point. A number cannot be interpreted as being any more accurate than its least significant digit, nor should a quantity be specified with any more digits than are justifiable by its measured accuracy. The numbers 128.1, 0.50, and 5.4, for example, imply quantities that have known accuracies of ± 0.1, ± 0.01, and ± 0.1, respectively, but the first is specified to four significant figures, while the second and third are specified to only two. If trailing zeros are placed after the decimal point, they carry the weight of significant figures. Thus, the number 0.50 means $0.5 \pm 0.01$.

The accuracy of any calculation can only be deemed as accurate as the *least* accurate number entering into the computation, and the number of significant figures that 0.50 can be claimed for the result should be set accordingly. For example, the product $128.1 \times 0.50 \times 5.4$ entered into a calculator produces the digits 345.87. But because 0.50 and 5.4 are specified to only two significant figures, the rounded-off result of the multiplication must be recorded as 350, also with two significant figures only. Note that a digit is rounded up if the digit to its right is 5 or more. If the digit to its right is less than 5, the digit is rounded down.

When numbers find their way into technical drawings, the number of significant figures takes on a special meaning. No part can ever be made to exact dimensions, because machine tools do not cut perfectly. A cutting tool wanders about its intended position during the machining process. Similarly, changes in temperature, humidity, or vibration during the cutting process can cause the tool to follow a less-than-perfect path. The tolerance of each dimension shown in the drawing specifies the degree of error that will be acceptable for the finished part. As a rule, creating parts with tight tolerances involves the use of more expensive machining equipment and more time, because material cuts must be made more slowly. These features add considerable expense to the finished part. As the designer, you must decide which dimensions are truly critical.

Suppose that you wish to make the vehicle chassis plate shown in Figure 4.4. The plate, to be made from 0.4-mm-thick aluminum, contains several threaded holes to which axle mounts are to be secured with screws. This fabrication job is a bit complicated for simple hand tools, hence you've decided to have a professional machinist make it. Indeed, such a job requires specialized machining tools, including a *milling machine*, *drill press*, and *reamer*. One issue that you might think about concerns the precision with which the part would need to be built. For a rough prototype, you might be content with an approximate version of the plate that can be produced quickly. For the finished product, you might want the machinist to take the extra time to adhere more closely to the specified dimensions. One way that engineers and machinists communicate on issues of this nature is through the numerical notations on parts drawings. Carefully note the labeled dimensions shown in Figure 4.4. These numbers communicate to the machinist the acceptable deviation, or *tolerance*, for each of the plate's various dimensions.

The numbers on the drawing in Figure 4.4 have precise meaning for any machinist who reads the tolerance table. In this case, only the hole diameters are especially critical. The length of the chassis plate, for example, is 25 cm. The numbers 25.0, 25.00, and 25.000, though all mathematically equivalent, would mean different things to the machinist. According to the tolerance table, the number 25.0, with one digit after the decimal point, should be interpreted by the machinist to mean $25 \pm 0.1$ cm. A chassis plate with a finished width diameter anywhere between 25.1 and 24.9 cm would be deemed acceptable. Similarly, the holes are specified as lying 15.00 cm apart, implying a machined tolerance of $15 \pm 0.05$ cm. The minimum and maximum tolerance limits for the hole centers as machined would be between 15.05 cm and 14.95 cm. According to the tolerance table, the most stringent dimensions of all are those of the hole diameters. Because these holes are intended to hold pins inserted by friction fit, their diameters are specified to three decimal points, implying a strict machining tolerance of $0.200 \pm 0.001$ cm.

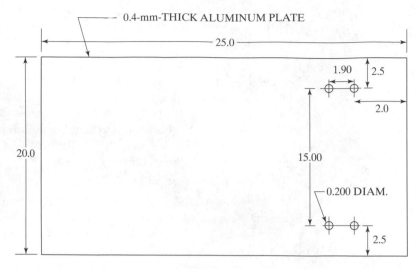

**MAIN CHASSIS PLATE**

| TOLERANCE TABLE | |
| --- | --- |
| All dimensions in cm | |
| X | ±0.5 |
| X.X | ±0.1 |
| X.XX | ±0.05 |
| X.XXX | ±0.001 |

**Figure 4.4.**   Main chassis plate with dimensions and tolerance table.

## PRACTICE!

1. Refer to the tolerance table on the logbook page shown in Figure 4.4. Compute the difference between the maximum and minimum permissible physical values for dimensions specified by the following numbers: 21.0 cm, 8.75 cm, 10 cm, 2.375 cm, 0.003 cm.

2. Write down the result of the following computation, using only the number of significant figures to which you are entitled:   $(45 + 8.2) \times 91.0 \div 12.1$.

3. What is the sum of the numbers $3.00 + 54.0 + 174 + 250$?

4. What is the sum of the integers $3 + 54 + 174 + 250$?

5. Using the tolerance table in Figure 4.4, write down the following dimensions specified to ±1 mm. 5.1 cm; 954 cm; 573 cm; 15 mm.

6. If all dimensions on a part are specified to be within ±1 mil (± 0.001 inch) what are the minimum and maximum angles between the sides of a 1-inch square?

## 4.3   PROTOTYPING AND BREADBOARDING

Designing anything for the first time requires careful planning and forethought. When a real physical device is being designed, the process should begin with a thorough consideration of all possible alternatives. An engineer usually resorts to estimation tools, sketches, and approximations to test each idea for fundamental feasibility. Later phases may involve computer simulations of the device, if appropriate. At some point in the design process, however, it will be time to construct a first working *prototype*. A prototype of a physical device is a mock-up of the finished product that embodies all its salient features but omits nonessential elements, such as a refined appearance or features not critical to the device's fundamental operation. Figure 4.5, for example, shows the prototype of a payload for a NASA rocket experiment.

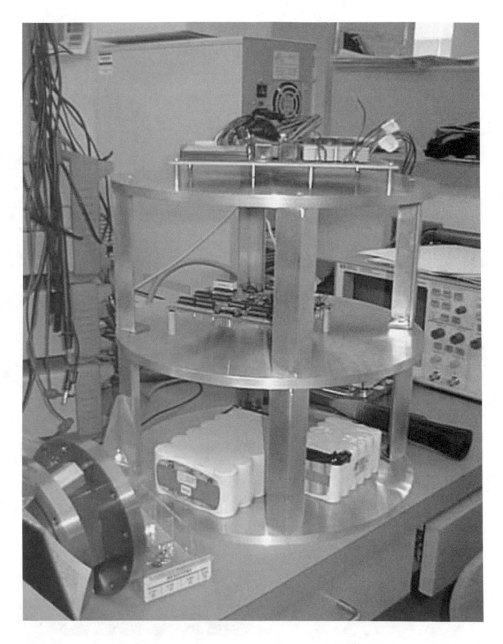

***Figure 4.5.***   Prototype of the payload for a NASA rocket experiment constructed at Boston University.

If the product is an electronic circuit, it can be built up on a *breadboard*. A breadboard allows an engineer to wire together the various components of a circuit, such as resistors, capacitors, and integrated circuits, by plugging them into holes aligned with spring-loaded clips located inside the breadboard. The clips make the electrical connections between components. In the prototype development stage, a breadboard readily permits changes and alterations to a circuit. A well-laid-out electronic breadboard is shown in Figure 4.6.

In production, the finished circuit is permanently soldered onto a printed wire board, such as those found inside computers, radios, and TVs. This form of circuit construction provides a durable, reliable finished product. The change in printed wire boards over the past several years has been dramatic as technology has evolved. Now

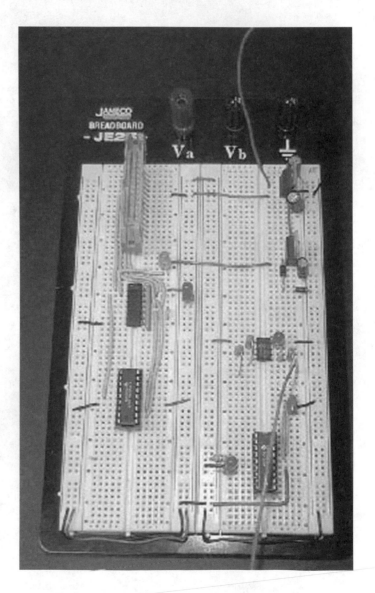

***Figure 4.6.*** A well-laid out circuit on an electronic breadboard.

integrated circuits capable of much more complex functions can be produced in extremely small packages that minimize interconnection wiring. Figures 4.7 and 4.8, for example, show state-of-the-art wire boards produced in 1995 and 2001, respectively. The first board uses many individual, dual-inline packaged, multipin integrated circuits, or "dip" chips, while the second uses much more compact and fully integrated surface-mount technology, where most of the operational function of the board is contained within the single integrated circuit seen in the lower left-hand corner of the picture. Note the reduced sizes of all components, including the integrated circuits, resistors, and capacitors, in the board of Figure 4.8.

Producing a mechanical product also requires the use of easy-to-modify prototypes. The latter should be fabricated from easily machined materials to produce a version that will enable testing and evaluation but may not be as durable or visually attractive as the finished product. A model car chassis plate, for example, while ultimately destined for fabrication from an aluminum plate, could be made from wood for initial tests of vehicle performance. Wood is easily drilled and formed and is readily available, but is much less durable than metal for demanding applications. Mechanical

***Figure 4.7.*** A printed wire board using "dip" chips produced around 1995.

prototypes also can be fabricated from various forms of angle iron and similar construction materials, as shown in Figure 4.9. The bars have holes in numerous places to allow for rapid construction and adaptation of prototypes.

Another example of physical prototyping can be found in the ball-and-socket human hip replacement joint of Figure 4.10. During development, this device might be fabricated from aluminum or stainless steel for testing in a cyclical loading machine. The finished product ready for implantation would be made from surgical-grade titanium and high-strength polymers at ten times the cost.

Engineers and architects who design large structures, such as buildings, bridges, and dams, face a handicap not encountered by other engineers. It's simply not practical to build a full-size prototype of these structures for testing purposes. It would be cost prohibitive, for example, to test the frame of a large sky scraper in, say, some remote desert location to determine its maximum sustainable wind speed before collapse. Engineers who build large structures rely on *scale modeling* to guide them through the prototyping phase. Scale modeling relies on dimensional similarity to extrapolate observations made on a model of reduced size to the full-sized structure. Effects such as structural stability, wind loading, temperature, and large-scale motion are readily scaled, tested, then extrapolated to the full-sized structure. Wind tunnels are widely used to test scale models for aerodynamic effects. Vibration, combustion, and wave phenomena do not scale well because these effects are governed by physical

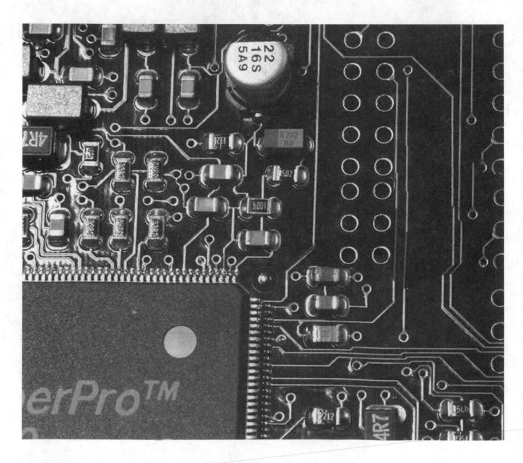

***Figure 4.8.*** A surface-mount printed wire board produced in 2001.

parameters that are fixed regardless of the frame of reference. (That's one reason why movie scenes filmed using scale models of ships at sea or burning buildings often look unrealistic.)

Software modules also undergo a prototype phase. Most software kernels (the part of a software program that does the actual thinking) are surrounded by a graphical user interface (GUI) that provides access to the user. A software designer will sometimes write and test the kernel portions of a program long before writing the graphical user interface.

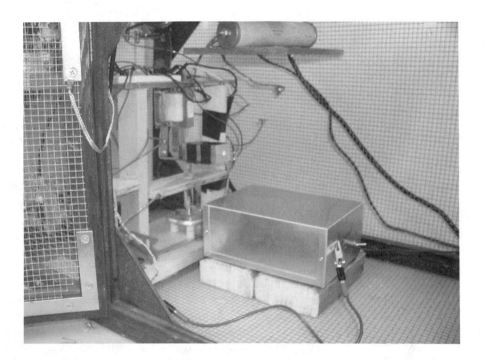

***Figure 4.9.***   Structure made from angle iron and a plywood base.

***Figure 4.10.***   Ball-and-socket human hip joint replacement. The finished product is made from expensive titanium. During development, initial prototypes might be made from aluminum or stainless steel.

**EXAMPLE 4.3: ELECTRICAL PROTOTYPING**

Prototyping is essential for the development and testing of electrical, mechanical, structural, and software devices. The prototype phase helps engineers reveal design flaws and problems that have escaped the initial planning and estimation phases. The following scenario illustrates the importance of careful prototyping in engineering design. In this scenario, two students, Juan and Tina, test a self-propelled model car they are designing for an in-class design competition. Tina has been in the process of testing a concept for a motor-timing circuit that will stop the vehicle after a prescribed time interval.

With Juan, her teammate, Tina first constructed a prototype of the car's chassis plate from plywood. Next, she mounted a small breadboard containing the car's electronic timing circuit and attached the drive motor to the chassis with some duct tape. She wired up the battery and connected it to the timing circuit. When she was finished with the wiring, she collected some diagnostic equipment: an oscilloscope to measure the output of the timing circuit, an ammeter to measure the motor current, and a voltmeter to monitor the battery voltage.

With the motor disconnected, Tina powered up the timing circuit and examined its output on the oscilloscope. Most of the circuit checked out, but Tina found a wire that she had forgotten to connect between two of the integrated circuits. After adding the wire, the circuit appeared to function as she had intended. It produced a voltage for driving the motor that lasted for about 7 seconds, the approximate time (according to her strategy) needed for the car to travel up the ramp specified in the contest rules.

Satisfied with the operation of the timing circuit, Tina next connected the motor to the load side of the circuit and turned it on. With the drive belt connecting the motor to the wheels disconnected, the motor turned nicely and stopped after 7 seconds, so she and Juan mounted the wheels, the gearbox, and the drive shaft onto the plywood frame and connected them to the motor with a small drive belt. The motor was about the size of a large D-cell battery. Tina held the car in midair. The motor again turned at a steady speed and cut out quickly when the timer circuit finished its timing pulse.

Tina next took the prototype and tried to run it up a large plank that served as a mockup of the competition ramp. The plank, purchased at a lumber store, had one end propped up on a chair to simulate an inclined ramp. After turning on the switch, the car prototype traveled about half a car's length and then stopped. At first, Tina was discouraged by this result. Was the timer faulty, or had she overlooked some fundamental design principle? She decided to perform further off-ramp tests of the vehicle. She clamped the car chassis in a vise with its wheels held in midair, connected the oscilloscope to the output of the timer, and connected the voltmeter across the battery and ammeter in series with the motor. When the switch was turned on, the motor turned as before for the entire seven-second duration of the timer signal. She noticed that the current to the motor was about 40 milliamperes and that the voltage on the 9-V battery dropped to about 8.2 V when the motor was running. Suspecting a battery drain problem, Tina next applied some friction to the wheels with her hand to simulate the load imposed by traveling up the ramp. The current to the motor jumped to 160 mA, and the battery voltage dropped to 3.9 V in response to the large current draw. This drop in voltage caused the timer circuit, which required a power source of at least 5 V, to cease functioning. Tina had been overloading the battery! She and Juan redesigned the layout of the chassis to accommodate a set of six 1.5-V AA batteries connected in series instead of a single 9-V battery. The former have a much higher current capability than the latter. The battery modification was easy to make on the wooden mockup of the chassis frame. At the expense of some added weight, the additional battery capacity would provide up to 200 mA of current with a voltage drop of only a volt or two. Tina's and Juan's battery performance problem was detected in the prototype phase.

The materials needed for creating prototypes can be found in many places. For building mechanical structures, no better place exists than your local hardware store. A well-stocked hardware store sells all sorts of nuts, bolts, rods, dowels, fasteners, springs, hinges, and nails. Many common electrical parts, such as wires, terminal connectors, tape, and switches also can be found at the local hardware store. A home center — a large hardware store, lumber yard, plumbing supply, electrical supply, and garden shop all rolled into one — is an excellent source of structural prototyping materials such as plywood, angle iron, pipe, and brackets (see, for example, *www.homedepot.com*).

A selection of very basic electronic parts and breadboards can be found at local consumer stores such as Radio Shack™. More complete selections at lower prices can be found from vendors on the Web. Examples include *www.digikey.com*, *www.jameco.com*, *www.abra-electronics.com*, and *www.newark.com*.

**EXAMPLE 4.4: CONSTRUCTING A PHYSICAL PROTOTYPE OF A MECHANICAL DRIVE**

Prism Corporation is a Boston-based company that is developing a process for producing inexpensive, environmentally friendly stampers for making compact disks (CDs.) A CD stamper is the master disk that is put into an injection molding machine so that common plastic versions of the CD can be mass produced. (While many computers now come with CD-write drives, the latter are not suitable for mass production of CDs. Compact disks that must be reproduced in large numbers are still made using CD stamper disks and injection molding machines.)

The present method for making CD stamper disks involves a time-consuming and expensive process in which a master disk is coated with a thin film of nickel metal. This coating is engraved with pits by a high-intensity laser, then peeled off and mounted on the stamper disk plate. The pits in the nickel coating represent the binary **1**s and **0**s of the digital information stored on the CD. The chemical substances used in processing the nickel coating eventually turn into hazardous waste that must be properly disposed of, again at significant cost. Prism has developed a process by which the CD imprint can be made directly on the stamper disk using a process known as *ion machining*. A nickel stamper disk is first covered with a light-sensitive photoresist, then exposed to a finely-focused laser beam as the disk spins. The laser is pulsed on and off in the binary sequence of information to be stored on the CD, thereby exposing the photoresist when the beam is on. When the photoresist is developed, it exposes the underlying stamper disk wherever a "pit" is to be etched. The disk is then put in a vacuum chamber and exposed to high-energy argon ions that etch the pits wherever the photoresist has exposed the underlying stamper disk. After etching, the rest of the photoresist layer is removed from the support disk, leaving the finished stamper.

One problem encountered by Prism is the need to hold and release the spinning stamper disk during the laser writing operation without physically touching the top surface of the disk. The disk must be held on a rotating *chuck* that is supported on high-speed, high-tolerance, noncontacting *air bearings*. The latter are similar to ball bearings but use a thin layer of high velocity air, forced between the rotating and stationary bearing surfaces of the chuck, to support the weight of the chuck. The thin air layer reduces vibrations significantly – a necessary feature if the pits are to be produced accurately on the CD stamper. Because the chuck is supported by air bearings, conventional vacuum techniques for holding objects without top-surface contact will not work, because there is no way to bring vacuum lines to the rotating chuck through the noncontact air bearing.

Prism has approached a company called Applied Electrostatics to develop an *electrostatic* method for holding the stamper disk on the rotating chuck. In this method, depicted in Figure 4.11, a sharp needle is held near the edge of the stamper disk and energized to a large positive voltage (several kilovolts.) The applied voltage produces an intense electric field at the needle tip, causing the needle to produce positive ions of air via a process known as *corona discharge*. These ions flow to the stamper disk which becomes positively charged, causing it to stick to the rotating chuck plate via electrostatic attraction (commonly called "static electricity" or "static cling"). After the laser writing process, the stamper is released from the chuck plate by applying a large negative voltage to the needle, thereby causing it to produce negative ions that neutralize the stamper disk and eliminate the static cling.

In order to design and test this electrostatic chuck concept, Applied Electrostatics has constructed the prototype shown in Figure 4.12. This prototype in no way resembles the actual electrostatic chuck that will become part of Prism's disk-stamper production system. Its sole purpose is to test the feasibility of using electrostatic forces in this

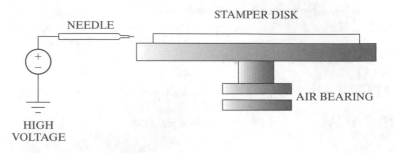

**Figure 4.11.** Electrostatic method for holding CD stamper plate on rotating air chuck.

**Figure 4.12.** Prototype of the electrostatic chuck developed for Prism Corp.

application. Key structural details that have found their way into the prototype include a simulated chuck-bearing plate made from aluminum, needle electrodes made from common sewing needles, high-voltage power supplies, and a method for tilting and spinning the mock-up of the bearing plate. The prototype is ugly, but functional. It has allowed Applied Electrostatics to test the salient features of the electrostatic chuck and to cycle through several design revisions as it develops its final product for Prism.

## 4.4 REVERSE ENGINEERING

Reverse engineering refers to the process by which an engineer dissects someone else's product to learn how it works. This design tool is particularly useful if your goal is to duplicate a competitor's product or create a similar one using your own technology. Reverse engineering is practiced on a regular basis by companies worldwide. Although it may appear to be an unfair practice, in reality it can be a good way to avoid patent infringement and other legal problems by specifically avoiding an approach taken by a competitor. Reverse engineering one of your own company's products can be a good way to understand its operation if its documentation trail has been lost or is inadequate. Reverse engineering in the software realm is encouraged in the writing of Web pages on the Internet. All of the major Web browsers provide a means to view and decipher the hypertext mark-up language (HTML) code, the instructions used to encode the Web page that has been downloaded into your computer. This practice fosters an environment of open information exchange that has been the hallmark of the Internet since its inception. In contrast, other forms of software written in high-level programming languages such as C, C++, Java, MATLAB™, Mathematica™, MathCad™, or Fortran, can be especially difficult to reverse engineer, particularly if the software has been poorly documented. A multitude of flow paths and logical junctions in software programs can lead to confusion on the part of the reader and make it hard to understand how a program operates.

## 4.5 COMPUTER ANALYSES

When engineers design something for the first time, they often build simplified prototypes to test the basic operating principles of the device. In many cases, simple hand calculations are all that are needed as a prelude to building a working prototype. At other times, however, more complex calculations are required to verify the feasibility of a design concept. Computers can be an important part of this verification process. *Simulation* is one major task that computers perform extremely well, and numerous software programs exist that help engineers simulate everything from bridges to electronic circuits. Examples of popular simulation programs include PSpice (electrical and electronic circuits), Pro-Engineer™ and Solidworks™ (solid modeling), Simulink™ (engineering systems), Ansoft™ (structural and field analysis), and Supreme™ (semiconductor devices).

General mathematical programs, such as MATLAB™, Mathcad™, and Mathematica™ are also extremely useful in the analysis of engineering problems. In the example to follow, a computer is used to perform an analysis related to a design competition. The analysis steps are illustrated using the MATLAB programming language. Although this specific software environment has been chosen for the example, the methodology illustrated is universal and could be used with any computer language or software program capable of performing numerical calculations. Other math packages, including Mathcad, Mathematica, or the computer languages C, C++, JAVA, or Fortran, for example, could be used with equal ease.

| | |
|---|---|
| **EXAMPLE 4.5: CALCULATING THE TRAJECTORY OF A FLYING HARPOON** | This example describes the efforts of two students, Tina and Juan, who have entered an engineering design competition called "Peak Performance." The objective of the competition is to build a small, self-propelled vehicle capable of climbing one side of a two-sided ramp within a 15-second time interval while an opposing vehicle attempts to climb the other side. The vehicle closest to the top of the ramp after 15 seconds is declared the winner. |

During a previous brainstorming session, Tina came up with the idea of launching a barbed harpoon over the top of the ramp in the path of the opposing vehicle as a way of blocking its path to the top. In this example, the two students explore their flying harpoon idea in more detail to assess its feasibility. Tina and Juan have wisely decided to build a prototype before finalizing their design. The prototype will consist of a launch tube, retractable rubber band, and harpoon that has a notched trailing end, as illustrated in Figure 4.13. One approach to building the prototype would be to throw together something that "seemed about right" from whatever rubber bands, rods, and tubes might be lying around. The students could then test the prototype to see if it worked and fiddle with it to optimize its performance. This method certainly would draw upon their engineering intuition and any previous practical experience in the construction of projectiles. Indeed, it probably would be fun to march ahead and try to build a working harpoon without first performing any kind of analysis. But a better approach, and one more likely to succeed, would be to test the feasibility of the launching mechanism first by simulating its operation on a computer. By so doing, the best values for key design parameters could be determined before actual construction. Parameters to be set might include the weight, diameter, and length of the harpoon, the force constant and retraction range of the rubber band, and the angle of inclination of the launch tube. Other constraints include the dimensions of the ramp (fixed by the design competition rules) and the maximum allowed size of the vehicle (30 cm × 30 cm × 30 cm). These last dimensions determine the maximum length of the harpoon which must fit within the allowed boundaries of the vehicle.

One possible trajectory for the harpoon is illustrated in Figure 4.14. Juan recalls from his physics class that if aerodynamic forces are negligible, the harpoon will follow the trajectory of a parabola. The choice for the harpoon's landing point, however, is arbitrary and becomes part of the students' offensive strategy. If they aim for a landing spot that lies past, but too close, to the top of the ramp, their opponent may be able to push

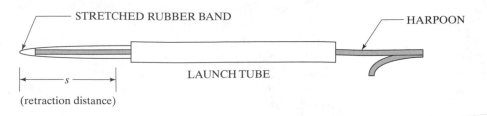

**Figure 4.13.**   Launch tube, stretched rubber band, and harpoon for Peak Performance offensive strategy.

the harpoon the short distance needed to reach the top of the ramp. If the harpoon's landing spot is over but too far down the opposite ramp, their opponent may be able to travel past the landing spot before the harpoon can be launched. Also, the farther away the landing spot is from the launch point, the greater the chance that small lateral (sideways) errors in launch angle will be amplified downstream over the harpoon's trajectory. These lateral errors might cause the harpoon to go over the edge of the ramp and land on the floor.

A computer can easily help Tina and Juan solve for the harpoon's trajectory. Variables of relevance include the harpoon's weight, the initial energy stored in the rubber band, and the angle of inclination of the launching tube. Before writing a computer program, however, they realize that they must understand the basic physics governing the harpoon's trajectory so that they can correctly code the program. Assuming that the rubber band, when released, will impart an initial velocity $v_o$ to the harpoon, the harpoon's equation of motion after ejection will be determined by Newton's law of motion:

$$\mathbf{F} = m\mathbf{a}. \tag{4-10}$$

Here, $\mathbf{F}$ is the total force on the harpoon, and $\mathbf{a}$ is the harpoon's acceleration. In this case, $\mathbf{F}$ acts in the $y$-direction only; hence, to the extent that aerodynamic forces can be ignored, the $x$- and $y$-components of Newton's law can be integrated to yield

$$x = x_o + v_{xo}t \tag{4-11}$$

and

$$y = y_o + v_{yo}t - \frac{mgt^2}{2}. \tag{4-12}$$

Here, $x_o$ and $y_o$ define the position of the center of gravity of the harpoon at $t = 0$, and $v_{xo}$ and $v_{yo}$ describe the initial $x$- and $y$-components, respectively, of the harpoon's initial velocity.[1] The quantity $m$ is the mass of the harpoon, and $g$ is the gravitational constant. The product $mg$ is the magnitude of the vertical force due to gravity. Note that gravity does not affect the value of $x$; the harpoon's horizontal velocity is constant in time.

The harpoon's exit velocity at $t = 0$ will be determined by the potential energy stored in the stretched rubber band. The latter can be computed from the band's force equation. For a rubber band exhibiting a linear restoring force, the force equation is given by

$$F_{band} = -ks. \tag{4-13}$$

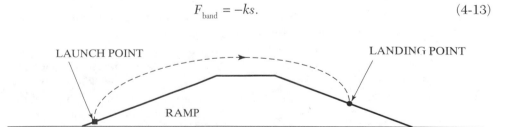

**Figure 4.14.**   Desired parabolic trajectory and landing site for the harpoon.

---

[1] When an object is propelled by an impulse force, it will acquire nearly instantaneously a velocity at t = 0. The horizontal component of velocity will remain fixed, while the vertical component will be subject to an acceleration due to gravity. This effect leads to a parabolic trajectory.

Here, $F_{band}$ is the force produced by the stretched band, $k$ is the band's spring constant in newtons per meter (N/m), and the stretch $s$ is the length that the band has been elongated from its unstretched position. The potential energy $E_P$ stored in the stretched band can be found by integrating the force equation as a function of $s$, yielding

$$E_P = -\int F\, ds = \int ks\, ds = \frac{ks^2}{2}. \tag{4-14}$$

When the harpoon exits the launch tube, all of the potential energy stored in the rubber band will be converted to kinetic energy of the moving harpoon (assuming that frictional losses in the launch mechanism are negligible). The kinetic energy of the harpoon can be expressed as

$$E_K = \frac{mv_o^2}{2}, \tag{4-15}$$

where $v_o$ is the harpoon's initial velocity. Equating $E_P$ to $E_K$ results in an expression for $v_o$ as a function of the band retraction $s$:

$$v_o = \sqrt{\frac{k}{m}}\, s. \tag{4-16}$$

The horizontal and vertical components of this exit velocity can be expressed by

$$v_{xo} = \sqrt{\frac{k}{m}}\, s\, \cos\theta \tag{4-17}$$

and

$$v_{yo} = \sqrt{\frac{k}{m}}\, s\, \sin\theta, \tag{4-18}$$

where $\theta$ is the angle of inclination of the launch tube. Using Eqs. (4-11), (4-12), (4-17), and (4-18), one can calculate the evolution of the trajectory $(x, y)$ in closed algebraic form to find the exact landing spot on the ramp. Given the complicated shape of the ramp, such a calculation would be very tedious to perform by hand. An alternative method is simply to use a computer to plot the harpoon's trajectory and to observe where it lands on the ramp. Although this second method will not provide as accurate an answer as the first one, it has visual appeal and is well suited for implementation on a computer.

Before any simulation can be performed, Tina and Juan must determine the various fixed parameters of the harpoon and the launching system. They've selected a weight of 100 gm (0.1 kg) for the harpoon (about the weight of one hundred paper clips). This choice represents an arbitrary starting point that can be changed later if necessary. The students next determine the spring constant of their rubber band. Several methods exist for measuring $k$. Tina and Juan decide to hang weights of various known values on the rubber band and measure its elongation, as shown in Figure 4.15. This simple experiment produces the data shown in Table 4-1. Of particular interest are the ratios of the vertical force $F$ to the displacement $y$. These ratios are equivalent to the spring constant of the rubber band, because $k = -F/y$. Tina and Juan rightfully ignore the values produced by the largest weights, because the rubber band is stretched beyond its elastic limit for these very large excursions. The values in Table 4-1 lying below the elastic limit yield an average *k* value of about 75 N/m. The data also suggest that Tina and Juan should confine their stretching excursions of the rubber band to about 10 cm or less in order to stay within the elastic limit. At this point, their calculation parameters consist of the values shown in Table 4-2.

**TABLE 4-1**   Force-Displacement Test Data for Rubber Band

| WEIGHT (kG) | FORCE* F (NEWTONS) | STRETCH y (CM) | F/y |
|---|---|---|---|
| 0.5 | 4.9 | 0.66 | 745 |
| 1.0 | 9.8 | 1.3 | 754 |
| 1.5 | 14.7 | 2.0 | 735 |
| 2.0 | 19.6 | 2.6 | 754 |
| 2.5 | 24.5 | 3.3 | 742 |
| 3.0 | 29.4 | 3.5 | 754 |
| 3.5 | 34.3 | 4.5 | 762 |
| 4.0 | 39.2 | 5.2 | 754 |
| 5.0 | 49.0 | 6.4 | 766 |
| 6.0 | 57.8 | 7.8 | 741 |
| 7.0 | 68.6 | 9.2 | 746 |
| 8.0 | 78.4 | 10.0 | 784 ** |
| 9.0 | 88.2 | 10.1 | 873 |
| 10.0 | 98.0 | 10.2 | 961 |

* $F = mg$, where $m$ is the mass and $g = 9.8$ N/kg is the gravitational constant.
** The probable elastic limit lies just above this point.

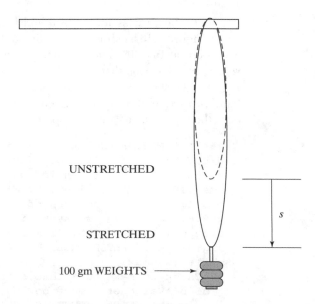

UNSTRETCHED

*s*

STRETCHED

100 gm WEIGHTS ⟶

**Figure 4.15.**   Measuring the elastic constant *k* of a rubber band.

**TABLE 4-2** Calculation Parameters for Numerical Simulation

| | |
|---|---|
| $k = 750$ N/m | elastic constant of rubber band |
| $m = 0.1$ kg | mass of harpoon |
| $g = 9.8$ N/kg | gravitational acceleration |
| $s_{max} = 10$ cm | maximum retraction distance for rubber band |

### 4.5.1 Plotting Trajectories Using MATLAB

MATLAB is a versatile and comprehensive programming environment particularly suited for engineering problems. MATLAB is similar in syntax to the programming language C, but it also provides commands that enable programmers to plot and organize data, manipulate matrices, observe variables, solve systems of linear equations, and solve differential equations. The strength of MATLAB lies in its ability to manipulate large amounts of data while allowing those data to be plotted and graphed. Tips on getting started in MATLAB can be found in any of several references. For example, one of the modules of the engineering series published by Prentice Hall (of which this book is a member) outlines the use of MATLAB in considerable detail.

In this example, a program that implements Juan and Tina's simulation using MATLAB is illustrated. Their trajectory program also could be written in any of several programming languages including C, C++, Java, Fortran, MathCad, or Mathematica.

One *possible* version of the code in MATLAB is shown below. Comment lines are preceded by percent signs (%). The program first plots a proportionately correct side view of the ramp, then prompts the user for values of the launch tube inclination angle and the amount of stretch of the rubber band. The program increments $t$ by $dt$ for each pass through the `while` loop. The looping continues as long as $x$ and $y$ remain within the boundaries $y > 0$ and $x < H$. After each position calculation, the program extends the trajectory plot from the most recently calculated $(x,y)$ position to the newly calculated position. Visual inspection of the plot reveals whether the trajectory hits the ramp at the desired location. This program was, in fact, the one used to produce the plots shown in Figure 4.16.

The best values for angle $\theta$ and band stretch distance $s$ are found by trial and error. Figure 4.16 illustrates a sampling of trajectories for the test values $s = 6$ cm, 7 cm, and 8 cm at $\theta = 40°$. The simulation indicates that no value of $s$ at this launch angle results in a harpoon that both clears the top of the ramp and lands on the other side. Figure 4.17 shows the same simulation at a launch angle of 70°. In this case, a rubber band retraction of 8 cm and launch angle of 70° lands the harpoon in a good spot: just over the top of the hill on the opponent's side of the ramp.

```
%%%%%    MATLAB PROGRAM CODE   %%%%%%%%
H=5;V=2;            %set boundaries of the problem in meters
k=750;             %elastic constant of rubber band in N/m
smax=10;            %maximum allowed s in cm
m=0.1;              %mass of harpoon in kg
g=9.8;             %gravitational acceleration in m/s2
s=input('ENTER AMOUNT OF RUBBER BAND STRETCH IN cm: ');
angle=input('ENTER ANGLE OF LAUNCH in DEGREES: ');
while (s>smax)
  %Check to make sure that s is less than smax
  disp('Value of s must be less than 7 cm');
  s=input('ENTER AMOUNT OF RUBBER BAND STRETCH IN cm: ');
end
```

```
while (angle > 90 & angle < 0)
  %Check to make sure that 0 < angle < 90
  disp('Angle must be less than 90 and greater than zero');
  angle=input('ENTER ANGLE OF LAUNCH in DEGREES: ');
end
%SET AXES FOR MAKING PLOT ON SCREEN:
axis([0 H 0 H]);
hold on
%DRAW RAMP ON THE SCREEN AS A DOTTED LINE:
plot([0 2.22 2.52 4.75],[0 1.02 1.02 0], '--');
%PREPARE TO CALCULATE AND PLOT
s=s/100;           %convert stretch from cm to meters
angle=angle*pi/180;%convert angle to radians
vo=sqrt(k/m)*s; %compute initial velocity
vox=vo*cos(angle);%compute x-component of initial velocity
voy=vo*sin(angle);%compute y-component of initial velocity
dt=H/(vox*100); %specify a time increment
xo=0.001; x=xo        %set initial value of x
yo=0.001; y=yo        %set initial value of y
t=0                   %set initial time to zero
%----------------------------------------------------------
while (y>0 & x<H)%calculate until trajectory goes out of bounds
  t = t+dt %increment the time
  xnew = xo + vox*t %compute new position x
  ynew = yo + voy*t - 0.5*g*(t^2); %compute new position y
plot ( [x xnew] , [y ynew] ); %plot latest segment of trajectory
  drawnow; %put segment on the screen
x=xnew; y=ynew %update values of x and y
end
```

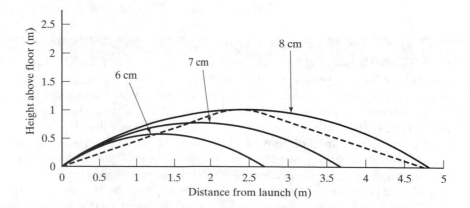

***Figure 4.16.*** Several calculated trajectories for rubber band elongation values $s = 6, 7,$ and 8 cm and launch angle $\theta = 40°$.

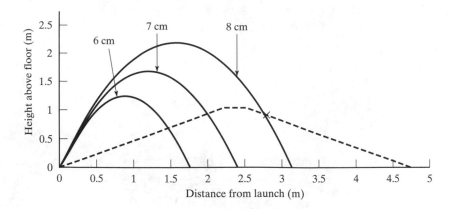

**Figure 4.17.** Repeat of simulation at a launch angle of $\theta = 70°$.

## PRACTICE!

1. Compute by hand the trajectory of Figure 4.16 for a 40° launch angle and rubber-band elongation of 8 cm. Verify that the projectile lands about 5 meters from its starting point on the other side of the ramp.

2. A rubber band has an elastic constant of 100 N/m. How much force will be required to elongate the band by 5 cm?

3. What is the potential energy stored in a spring that has a restoring force of 1 kN/m and has been stretched by 1 cm?

4. What is the potential energy stored in a spring that has a restoring force of 500 N/mm and has been compressed by 5 mm?

5. Compute the initial velocity of a 100-gm projectile that is launched by stretching a rubber band by 10 cm. The rubber band has an elastic constant of 50 N/m.

6. What launch angle will result in a harpoon that travels the farthest distance in Example 4.5. Assume that the harpoon's launch velocity and angle will cause it to travel over the top of the ramp and land somewhere on the other side.

7. Draw the flowchart of a program designed to compute the path of a bowling ball after it leaves a bowler's hand.

8. Draw the flowchart of a program designed to compute the height of a helium-filled balloon after it has been released from sea level.

### PROFESSIONAL SUCCESS: THE ROLE OF COMPUTERS IN SOCIETY

Computers have become so enmeshed in our lives that it's hard to imagine what society was like without them. Computers are used in business, commerce, government, education, finance, medicine, avionics, social service, and, of course, engineering. Computers have even become a form of recreation. One can debate the merit of computers and their effect on human relationships (Is an AOL chat better than a phone conversation?), but in the world of engineering, computers are indispensable. Like most people, engineers use computers for communication, information retrieval, data processing, word processing, electronic mail, and Web browsing. But the true worth of the computer to the engineer lies in its ability to perform calculations extremely rapidly. A computer can be programmed to perform all sorts of numerical calculations. Commercial programs for simulation, spreadshects, and graphing enable engineers to determine everything from the stresses on mechanical parts and the operation of complicated electronic circuits to the force loads on building frames and theoretical predictions of rocket launches. The availability of the computer and the abundance of software tools greatly enhance the productivity of engineers in all disciplines.

**EXAMPLE 4.6:**
**METHOD OF**
**NUMERICAL**
**ITERATION BY**
**COMPUTER**

The field of micro-electromechanical systems, or MEMS, has become increasingly important over the past several years. MEMS devices are tiny micro-scale machines made from silicon, metals, or other materials. They are fabricated using tools borrowed from integrated-circuit manufacturing: photolithography, pattern masking, deposition, and etching. MEMS devices are beginning to find their way into mainstream engineering design solutions. The sensors used to deploy safety airbags in most automobiles, for example, are built around tiny MEMS accelerometers that are on the order of a square millimeter in size.

One technique for fabricating MEMS devices is called *surface micromachining*. The basic steps involved in surface micromachining are shown in Figure 4.18. A silicon substrate is patterned with alternating layers of polysilicon and oxide thin films that are used to build up a desired mechanical structure. The oxide films serve as *sacrificial layers* that support the polysilicon layers during fabrication but are removed in the final steps of fabrication. This construction technique is analogous to the way that arches of stone buildings were made in ancient times. Sand was used to support stone pieces and was removed when the building could support itself, leaving the finished structure.

One simple MEMS device used in numerous applications is shown in Figure 4.19. This *double-cantilevered actuator* consists of a bridge supported on two ends and situated over an underlying, fixed activation electrode. A side view is shown in Figure 4.19. The bridge has the shape of a square when viewed from the top. The bridge sits atop an insulating layer and silicon substrate. When a voltage is applied between the bridge and the substrate, the electrostatic force of attraction causes the bridge to deflect downward. This vertical motion can be used to move other parts or to perform useful functions. For example, the deflecting bridge can open and close tiny valves, change the direction of reflected light, pump fluids, or mix chemicals in small micromixing chambers. A MEMS designer must know the relationship between the voltage applied to the bridge and its deflection. For a given applied voltage $V$, the electrostatic force will be given approximately by the following equation:

$$F_e = \frac{\varepsilon_0 A V^2}{2(g-y)^2}. \qquad (4\text{-}19)$$

Here, $y$ is the bridge deflection, $A$ is the area of the bridge as seen from the top, and $g$ is the gap spacing between the bridge and the electrode at zero deflection. The *permittivity* constant $\varepsilon_0$ is equal to $8.85 \times 10^{-12}$ F/m for air (farads per meter). Note that

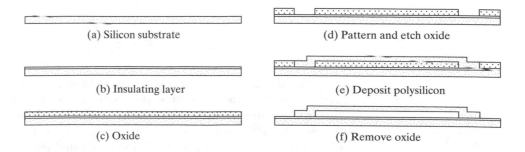

(a) Silicon substrate        (d) Pattern and etch oxide

(b) Insulating layer        (e) Deposit polysilicon

(c) Oxide        (f) Remove oxide

**Figure 4.18.** Fabrication sequence for a simple MEMS actuator: (a) Begin with a silicon wafer substrate; (b) deposit an insulating layer of silicon nitride; (c) deposit a thick layer of silicon dioxide ("oxide"); (d) pattern and etch the oxide using photolithography and masking techniques; (e) deposit a layer of polysilicon that will become the actuator bridge; (f) remove the oxide, leaving an air gap between the actuator and the substrate.

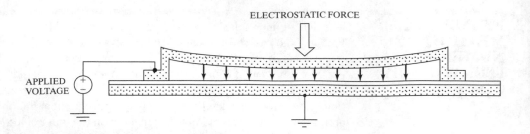

**Figure 4.19.**   Applying a voltage between the actuator and the substrate causes deflection of the actuator bridge. This mechanical motion can be used to move other devices, change the direction of reflected light, pump liquids and gases, or perform other operations on a microscopic scale.

the electrostatic force increases with increasing deflection and becomes infinite for $y = g$ (i.e., when the net gap spacing becomes zero). This increase in force with deflection would cause the bridge to collapse completely when any voltage was applied were it not for the counteracting mechanical *restoring force* of the elastic polysilicon used to make the bridge. To first order, the restoring force will be proportional to the bridge deflection and can be expressed by the simple equation

$$F_m = -ky. \tag{4-20}$$

This force is exactly analogous to that of the rubber band used to power the harpoon in Example 4.5, in that it increases in proportion to the deflection. In the case of the MEMS device of Figure 4.19, the mechanical restoring force prevents the bridge from collapsing completely when a voltage is applied. As the deflection increases, the restoring force increases also. At some value of deflection, the mechanical force becomes equal to the electrostatic force, allowing no further deflection. A MEMS designer is extremely interested in this equilibrium point, because it determines the bridge deflection for a given applied voltage.

The equilibrium deflection point $y$ that results from a given applied voltage is difficult to determine from hand calculations only. In principle, one can find it by equating the magnitudes of the electrostatic and mechanical forces given by Eqs. (4-19) and (4-20), yielding the following cubic equation:

$$ky = \frac{\varepsilon_0 A V^2}{2(g-y)^2}. \tag{4-21}$$

Solving this force balance equation by hand is difficult. (Try it!) The calculation is well suited, however, for solution on a computer using the method of numerical *iteration*. In this latter method, the computer tries many different values of $y$ until it finds one for which both sides of Eq. (4-21) match. One can instruct the computer, for example, to begin with some very small value of $y$ and then increase it by small amounts until the solution point is found. This iterative method is ideal for implementation on a computer, because it typically involves many repetitive calculations that would be time-consuming if performed by hand.

The flowchart of Figure 4.20 illustrates the steps needed to find the equilibrium point by the method of iteration. The program begins with a small value of $y$, then successively increases $y$ by $dy$ until the left-hand and right-hand sides of Eq. (4-21) agree to within some residually small value set by the programmer.

The circles in Figure 4.21 show the results of the computation for several values of applied voltage and the parameters listed in Table 4-3. The dotted curve shows the complete analytical solution for comparison. The latter was plotted by solving Eq. (4-21) for numerous values of $y$ and V, storing the data points, and then using the "plot" command in MATLAB to graph the data points as a smooth curve.

Above about 45 volts, the restoring force is no longer capable of holding back the electrostatic force, and the deflection becomes "infinite," i.e., the bridge collapses all the way to its underlying electrode. This phenomenon is called *snap through* in the world of MEMS. Snap through commonly occurs at a deflection of about one-third of the zero-voltage gap spacing.

The iteration leading to Figure 4.21 can be performed using any number of available software programs. Sample program code listings written in C++ and MATLAB that

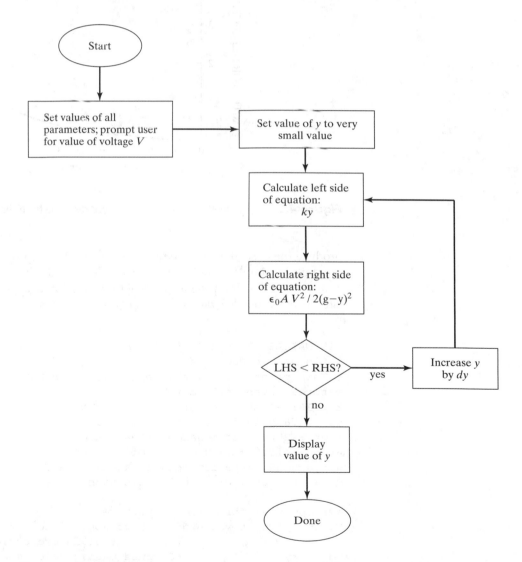

**Figure 4.20.**  Flowchart for the iterative solution of Eq. (4-21).

**TABLE 4-3** Parameters Used in the Iterative Calculations

| QUANTITY | VARIABLE | VALUE |
|---|---|---|
| Restoring force of polysilicon structure | $k$ | 30 N/m |
| Size of bridge actuator | $side$ | 250 μm |
| Gap spacing | $gap$ | 5 μm |
| Permittivity of air | $\varepsilon_0$ | $8.85 \times 10^{-12}$ F/m |

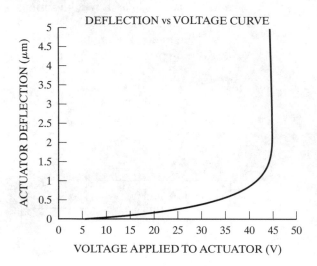

**Figure 4.21.** Deflection versus voltage curve resulting from the simulation of Fig. 4.20.

produce the desired results are provided below. The C++ program can be turned into a C program by simply changing the header files and the functions used for reading and writing variables. The programs prompt the user for an applied value of voltage, calculate the corresponding deflection $y$, and display the result.

```
%%% -----------------------------------------------------%%%
%%% MATLAB PROGRAM CODE LISTING %%
% Lines preceded by a percent sign (%) are comment lines
% This program finds the deflection of a MEMS bridge actuator
% for a given applied voltage.
k = 30;%elastic restoring force in N/m
eo = 8.85e-12;%electric permittivity of air
side = 250e-6;%dimension of each side of actuator
Area = side^2%compute area of actuator
gap = 5e-6; %spacing between actuator and activation electrode
dy = gap/100%incremental deflection to be used in iteration

%Prompt user for value of voltage:
V=input('Enter value of voltage applied to actuator:  ')
y=0;                            %initialize deflection to zero
Fm = k*y;              %Compute magnitude of mechanical force
Fe = eo*Area*V^2/2(gap-y)^2;%Compute magnitude of electrostatic force
```

```
%---------------------------------------------------------
while Fm < Fe;
    y = y + dy;              %Try a slightly larger deflection
Fm = k*y;                    %Recompute magnitude of mechanical force
Fe = eo*Area*V^2/2(gap-y)^2%Recompute magnitude of electrostatic force
end
% Display result:
disp('Deflection in microns when Fe = Fm: ' y*1e6)
%%% ----------------------------------------------------%%%
```

```
/*-------------------------------------------------------*/
/* C++ PROGRAM CODE LISTING */
/* Program to simulate MEMS actuator position versus voltage curve */
#include<stdio.h>
#include<math.h>
int main() {
        int k = 30;
        float eo = 8.85e-12,
             side = 250e-6,
             area,
             gap = 5e-6,
             dy;
        float v,
             y,
             Fm,
             Fe;
/*Square "side" to get area (raise side to the power 2) */
        area = pow(side,2);
        dy = gap/100;
/* Prompt user for value of voltage */
cout << "Enter value of voltage applied to actuator:  ";
cin >> v;
        y = 0;
        Fm = k * y;
        Fe = (eo * area * pow(v,2)) / pow(gap y,2);
        while(Fm < Fe) {
                y = y + dy;
                Fm = k * y;
                Fe = (eo * area * pow(v, 2)) / pow(gap y, 2);
        }
cout << "Deflection in microns when Fe=Fm: " << y*1e6 << endl;
        return 0;
}
/*-------------------------------------------------------*/
```

---

**PROFESSIONAL SUCCESS: GARBAGE IN, GARBAGE OUT**

The computer has become an indispensable tool for engineers. It can perform calculations much more rapidly than can a human. It's superb for storing and retrieving data and producing graphical images. But the computer is no substitute for thinking. A computer should be used to enhance the capabilities of an engineer, not replace them. A computer will faithfully follow its program code but is unable to pass judgment on the worth of the results. It's up to the engineer to provide the computer with information and program statements that are meaningful and relevant. If you program a computer to calculate the weight of structural steel for a bridge, it can do so flawlessly. But it never can tell you whether such a bridge is feasible, whether the stress-strain equations that it used in its program are correct, or whether the bridge should be built at all. Only an engineer with experience and good judgment can make those determinations.

When a computer crunches numbers that have little meaning due to programmer error, or when it operates with faulty data that was erroneously entered, the computer operates in a "garbage in, garbage out" (GIGO) mode. GIGO refers to a situation in which bad input data or bad program code lead to a computationally correct but meaningless output. A GIGO condition can be prevented by testing a program on simple, well-known examples for which the solution can be easily computed by hand. If the computer can provide correct answers to numerous simple problems, then it's probably programmed correctly and is ready to be used on more complex problems.

## 4.6    THE INTERNET

When the first edition of this book was written, the use of the Internet was described as "an up-and-coming engineering design tool" that was "rapidly gaining popularity." Today, it's virtually impossible to imagine engineering without the Internet. Indeed, the linking of millions of Internet host sites has become an important tool not only for engineers, but also for individuals involved in business, commerce, science, politics, education, and recreation. The availability of data sheets in HTML (hypertext markup language) or PDF (portable document format) has all but made the printed technical data book obsolete. Parts and supplies are now more easily purchased from on-line stores than they are from "brick-and-mortar" vendors. The first stop for any engineer seeking information about a particular type of product is likely to be an Internet search engine.

### 4.6.1    "I Saw It on the Internet. It Must be True."

While a comprehensive introduction to the Internet is beyond the mission of this text, a discussion of *when* to use the Internet as part of the engineering design process is certainly appropriate. When you access information from the Internet for an engineering project, be sure that it comes from a reliable source. Information is not necessarily accurate just because it has been posted on the Web. Information that comes from the Web sites of reputable companies or the government is most likely reliable. Information from personal student Web pages, student project sites, the amateur press, lone information providers, and other sources outside the mainstream should be viewed with more skepticism. Before the Internet, it took a great deal of time for rumors and misinformation to propagate around the technical community. Now this process takes minutes. Information from one Web site can be copied to another, leading to multiple propagation of errors. Although the Web has provided instant access to limitless information, its onset eliminated an important filter provided by print media: Web pages are inexpensive to produce; hence, almost anyone can produce them. Before the Web, only serious companies could afford to disseminate information to the general public. In the print age, technical information seemed more grounded in legitimacy. Now, worthless information can be indistinguishable from valued information obtained from serious providers. Choose your Internet sources with care.

One other word of caution: The World Wide Web has been around only since the early 1990s. Its growth has been explosive, but the information it can provide on a particular subject is only as complete as the time someone has taken to post it. A great body of engineering data and knowledge exists that was present long before the Web came into being. As a source of information, the Web can only augment the hundreds of millions of books, reports, and periodicals available in the world's libraries. Although the Web and its associated computer-based searching tools are important sources of information for your design project, they should not be the only sources.

**PRACTICE!**

1.  Use the search feature of your Web browser to perform a search on the term "oil platform." Record the number of listings. Now, narrow the search by adding, in succession, the following additional keywords: "photo," "north," "sea," and "terminal." Determine how many fewer listings are obtained each time the search is narrowed by an additional key word.

2.  Use the search feature of your Web browser to perform a search on the term "catapult." Record the number of listings. Now, narrow the search by adding, in succession, the following additional keywords: "photo," "medieval," "trebuchet," and "reenactment." Determine how many fewer listings are obtained each time the search is narrowed by an additional keyword.

3.  Use the search feature of your Web browser to perform a search on the term "space station." Record the number of listings. Now, narrow the search by adding, in succession, the following additional keywords: "photo," "NASA," "Russian," and "shuttle." Determine how many fewer listings are obtained each time the search is narrowed by an additional keyword.

## 4.7 SPREADSHEETS IN ENGINEERING DESIGN

A spreadsheet is a programmable table in which each cell consists of text, fixed numerical data, or a formula that relies on other numbers in the spreadsheet for its value. Popular spreadsheets include Microsoft Excel™, and Lotus 1-2-3™. Engineers may use spreadsheets in all phases of the design process for tasks such as planning budgets, tracking parts lists, analyzing results, and performing calculations. A spreadsheet becomes particularly useful when a problem is complex and has many interrelated variables. By programming a spreadsheet to model a complex problem, an engineer can see the effect that changing a single variable has on the entire system. The following two examples illustrate the usefulness of spreadsheets in engineering design. The spreadsheet tables shown are meant to reflect a generic form typical of most types of commercial spreadsheet software.

**EXAMPLE 4.7: CALCULATING THE CENTER OF MASS**

This examples describes one sphase in the design activities of two students, who have entered an engineering design competition called "Peak Performance." They are in the process of designing a small, battery-powered, wedge-shaped vehicle that must scale a 2.5-meter long ramp within a 15-second time interval. They have determined that their model car will perform best if the its center of mass is located midway between its front and rear axles. Their simulations have shown that if the center of mass is too far forward, the rear wheels will not maintain enough traction to drive the car up the ramp. Conversely, if the center of mass is too far toward the rear, the high torque of the motor and gears may cause the front of the car to lift off the ramp during startup or during contact with an opposing vehicle. The students now address the problem of where to mount the various components on the chassis. The location of these vehicle components will determine the location of the vehicle's center of mass.

Tina has weighed each component and has entered her list, shown in Figure 4.22, into her logbook. She's also entered the sketch of Figure 4.23 which outlines one possible layout for the components. She's calculated the center of mass of the can's wedged-shape frame from basic principles and has found that it lies about 7 cm behind the midpoint between the two axles.

The students' next step is to decide where to place each of the components on the vehicle chassis. Juan has prepared a spreadsheet, shown in Table 4-4, to calculate the center of mass of the entire vehicle, including all its components. (Note that Table 4-4 shows the contents of each cell in the spreadsheet, not what appears on Juan's computer screen.) The location $x$ of each part represents the position of its own individual center of mass relative to the center of mass of the chassis. The latter is located at the midpoint between the two axles. The moment $M_n$ of the $n^{th}$ part is equal to its position times its mass:

$$M_n = x_n \times m_n. \tag{4-22}$$

2/14/99
Measurements of vehicle parts

| ITEM | APPROX* WEIGHT [gm] | PROPOSED LOCATION |
|---|---|---|
| chassis | 1000 | [defines center line] |
| wedge fr. | 400 | −7 cm |
| battery | 120 | +8 |
| switch | 10 | +5 |
| motor | 200 | −7 |
| gearbox | 50 | −10 |
| launch tube | 15 | −12 |

*measured on pan balance

**Figure 4.22.**   Logbook page showing weights of key vehicle components.

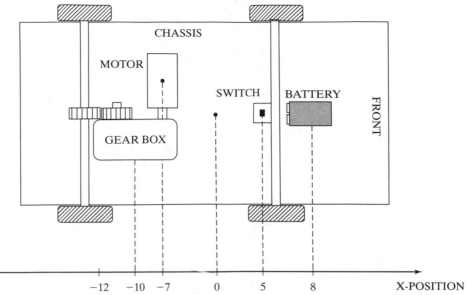

**Figure 4.23.**  Possible layout of vehicle components.

**TABLE 4-4**  Cell Entries in Spreadsheet that Calculates Center of Mass

|  | **A** | **B** | **C** | **D** |
|---|---|---|---|---|
| **1** | PART NAME | MASS m (gm) | LOCATION x (cm) | MOMENT m (gm*cm) |
| **2** | Chassis | 1000 | 0 | =B2*C2 |
| **3** | Wedge Frame | 400 | −7 | =B3*C3 |
| **4** | Battery | 120 | 8 | =B4*C4 |
| **5** | Switch | 10 | 5 | =B5*C5 |
| **6** | Motor | 200 | −7 | =B6*C6 |
| **7** | Gear Box | 50 | −10 | =B7*C7 |
| **8** | Launch Tube | 15 | −12 | =B8*C8 |
| **9** | **Total Weight** | =SUM(B2:B8) | | |
| **10** | **Total Moment** | | | =SUM(D2:D8) |
| **11** | **Net C.O.M.** | | =D11/B10 | |

The center of mass $x_{CM}$ of the entire ensemble of parts is given by the sum of the moments divided by the sum of the masses:

$$x_{CM} = \frac{\sum M_n}{\sum m_n}. \tag{4-23}$$

The output of Juan's spreadsheet, as it appears on his computer screen, is shown in Table 4-5. His calculations show that the center of mass of the vehicle (Net C.O.M. in the spreadsheet table) will lie at $x = -2.16$, or about 2 cm behind the geometrical center of the axle mounts. The force calculations that he and Tina have previously performed using a C++ program indicate that the center of mass needs to lie within 1 cm of the vehicle's midpoint.

The students discuss possible alternative arrangements for the vehicle components. Juan tries to simulate moving the battery forward by changing the entry for battery position on his spreadsheet. Because the battery is connected to the motor by wires of arbitrary length, this change is an easy one to implement. The change, however, shows that moving the battery all the way to the 15-cm forward position merely shifts the center of mass forward to the −1.7 cm position.

Tina suggests adding a counterweight somewhere between the midpoint of the vehicle and its front axle. Juan inserts a new row into the spreadsheet and enters data for the counterweight into the cells, as shown in Table 4-6. The students experiment with

**TABLE 4-5**  Screen Output of the Spreadsheet of Table 4-4

|  | A | B | C | D |
|---|---|---|---|---|
| **1** | **PART NAME** | **MASS m (gm)** | **LOCATION x (cm)** | **MOMENT m (gm*cm)** |
| 2 | Chassis | 1000 | 0 | 0 |
| 3 | Wedge Frame | 400 | −7 | −2800 |
| 4 | Battery | 120 | 8 | 960 |
| 5 | Switch | 10 | 5 | 50 |
| 6 | Motor | 200 | −7 | −1400 |
| 7 | Gear Box | 50 | −10 | −500 |
| 8 | Launch Tube | 15 | −12 | −180 |
| 9 | **Total Weight** | 1795 | | |
| 10 | **Total Moment** | | | −3870 |
| 11 | **Net C.O.M.** | | −2.16 | |

**TABLE 4-6**  Screen Output of Modified Spreadsheet

|  | A | B | C | D |
|---|---|---|---|---|
| **1** | **PART NAME** | **MASS m (gm)** | **LOCATION x (cm)** | **MOMENT m (gm*cm)** |
| 2 | Chassis | 1000 | 0 | 0 |
| 3 | Wedge Frame | 400 | −7.1 | −2840 |
| 4 | Battery | 120 | 8 | 960 |
| 5 | Switch | 10 | 5 | 50 |
| 6 | Motor | 200 | −7 | −1400 |
| 7 | Gear Box | 50 | −10 | −500 |
| 8 | Launch Tube | 15 | −12 | −180 |
| 9 | Counterweight | 200 | 10 | 2000 |
| 10 | **Total Weight** | 1995 | | |
| 11 | **Total Moment** | | | −1910 |
| 12 | **Net C.O.M.** | | −0.96 | |

different values for the location and mass of the counterweight. The spreadsheet allows them to track changes in the center of mass as it's modified by the counterweight. Competition rules specify a 2-kg vehicle weight limit for the entire vehicle, including batteries. The spreadsheet allows the students to see how close they come to this limit as the mass of the counterweight is increased. The cells in Table 4-6 show the results of placing a 200-gm counterweight 10 cm forward of the vehicle midpoint. The center of mass has been shifted to −0.96 cm behind the midpoint, which lies just on the edge of the targeted range of ±1 cm. The total weight has increased to 1,995 gm, which is five grams below the maximum 2-kg limit. Tina and Juan decide to stay with these parameters so that they will have a small safety margin with respect to weight, in case they need to add small nuts, bolts, glue, or tape during the competition.

**EXAMPLE 4.8: KEEPING TRACK OF COST WITH A SPREADSHEET**

A spreadsheet also helps Juan and Tina keep track of the total cost of their vehicle. Their professor has set a total cost limit of $100, including batteries, so that no team can produce a better vehicle simply by having unlimited funds. An accounting of the costs must be submitted by each participating team. Tina has set up a spreadsheet to track these costs. The output screen of the spreadsheet is shown in Table 4-7.

**TABLE 4-7**  Cost-Tracking Spreadsheet. The Unit Cost Column Indicates the Cost Per Item, and the Extended Column is Equal to Quantity × Unit Cost

|  | A | B | C | D | E |
|---|---|---|---|---|---|
| **1** | ITEM | QUANTITY | UNIT COST $ | | EXTENDED $ |
| **2** | Chassis | 1 | 12.50 | | 12.50 |
| **3** | Wedge Frame | 1 | 15.00 | | 15.00 |
| **4** | Battery | 12 | 2.19 | | 26.28 |
| **5** | Switch | 3 | 2.29 | | 6.87 |
| **6** | Motor | 1 | 3.49 | | 3.49 |
| **7** | Gear Box | 1 | 5.99 | | 5.99 |
| **8** | Launch Tube | 1 | 0.67 | | 0.67 |
| **9** | Counterweight | 1 | 0.85 | | 0.85 |
| **10** | Rubber bands | 12 | 0.04 | | 0.48 |
| **11** | 6-32 Screws | 24 | 0.06 | | 1.44 |
| **12** | 6-32 Nuts | 24 | 0.05 | | 1.20 |
| **13** | 6-32 Washers | 24 | 0.02 | | 0.48 |
| **14** | Two-Part Epoxy | 1 | 2.29 | | 2.29 |
| **15** | Wheels | 4 | 1.59 | | 6.36 |
| **16** | Spool of Wire | 1 | 2.49 | | 2.49 |
| **17** | Tape | 1 | 2.59 | | 2.59 |
| **18** | Metal Brackets | 6 | 1.19 | | 7.14 |
| **19** | Screw Thread | 1 | 0.99 | | 0.99 |
| **20** | Wing Nut | 2 | 0.25 | | 0.50 |
| **21** | **TOTAL** | | | | **$97.61** |

The *Unit Cost* column indicates the cost per item, while the *Quantity* column indicates the number of parts of that type that has gone into the vehicle. The *Extended* column, equal to the product *Quantity × Unit Cost*, shows the total cost for each part type. The TOTAL entry at the bottom provides the sum of all the extended prices. As each part is added to the vehicle, the students update their spreadsheet. If the total cost exceeds $100, they can use the spreadsheet to experiment with eliminating various parts, such as extra nuts and bolts, if the design allows it, until the total cost is in compliance. At first, they allot 16 batteries for use on competition day, but this quantity causes the total cost to exceed $100. By using the spreadsheet, they are able to determine by trial and error that reducing the battery allotment to 12 results in a total cost less than $100.

**PRACTICE!**

1. Verify the calculated cells in the spreadsheet of Table 4-6.
2. Find the center of mass in the *x-y* plane of a set of objects positioned as follows. The numbers in parentheses indicate each object's *x-y* coordinates: *Object 1*: 1.2 kg (0.2 m, 0.4 m); *object 2*: 3.3 kg (1.3 m, −2.3 m); *object 3*: 0.9 kg (0.8 m, 0.4 m); *object 4*: 0.2 kg (0 m, 1.7 m).
3. Using published population data, find the approximate geographical "center-of-mass" of the country in which you reside.

**EXAMPLE 4.9: MEMS ACTUATOR REVISITED**

The method of solution by computer iteration was illustrated in Example 4.6 using programs written in C⁺⁺ and MATLAB. In Table 4-8, this same problem is solved using an Excel™ spreadsheet formulation for the case $V = 35$ V. Each cell in column 1 represents a "test" value of $y$ and is greater than the value of $y$ in the cell above it by the incremental value of $dy$ that is entered into cell B6. When this variable is entered into cell formulas elsewhere in the spreadsheet, the Excel "dollar-sign" notation $B$6 is used, rather than the simple row-column notation B6, to make sure that any references to B6 will be preserved when cell contents are copied from one cell to another. Without the dollar sign notation, the cell coordinates B and 6 would be automatically incremented during copy operations. In fact, the dollar sign notation is used to refer to each of the fixed quantities under the B column of the spreadsheet.

Columns 2 and 3 in Table 4-8 contain the computed values of the left-hand and right-hand sides of Eq. 4.21, respectively. The entries in column 4, called the *residual* values, represent the difference between the column 2 and 3 entries. We seek the row for which the residual is zero, i.e., the value of $y$ at which the mechanical force $F_m$ balances the electrostatic force $F_e$. The actual output of this spreadsheet is shown in Table 4-9. No single value of $y$ shown in the table yields a precise residual of zero, but the values $y = 0.5$ μm and $y = 0.75$ μm bound the correct answer, because the residual in column 4 changes from a negative to a positive value between these two rows.

Table 4-10 shows a second spreadsheet, identical to the first, except that the upper and lower bounds of $y$ values tested are chosen as 5 μm and 6.1 μm, and the increment $dy$ is reduced to 0.01 μm, or about a tenth as large as the $dy$ used in Table 4-9. The results of this second spreadsheet show that the actual equilibrium point lies somewhere between y = 0.57 µm and y = 0.58 µm. If a still more precise answer is desired, then these *y* values could be set as new upper and lower bounds and *dy* reduced to an even smaller value.

**TABLE 4-8**  Spreadsheet Solution to the MEMS Actuator Problem of Example 4.6

|  | A | B | C | D | E | F | G |
|---|---|---|---|---|---|---|---|
| 1 | VOLTAGE: | 35 | VAL. NO. | COLUMN 1 | COLUMN 2 | COLUMN 3 | COLUMN 4 |
| 2 | k: | 30 |  | y | FM | FE | DIFFERENCE |
| 3 | gap: | =5°10^(-6) | 1 | 0 | =$B$2°D3 | =$B$7°$B$5°$B$1^2/($B$3 D3)^2 | =E3-F3 |
| 4 | side: | =250°10^(-6) | =C3+1 | =D3+$B$6 | =$B$2°D4 | =$B$7°$B$5°$B$1^2/($B$3 D4)^2 | =E4-F4 |
| 5 | area: | =(A4)^2 | =C4+1 | =D4+$B$6 | =$B$2°D5 | =$B$7°$B$5°$B$1^2/($B$3 D5)^2 | =E5-F5 |
| 6 | dy: | =B3/20 | =C5+1 | =D5+$B$6 | =$B$2°D6 | =$B$7°$B$5°$B$1^2/($B$3 D6)^2 | =E6-F6 |
| 7 | eo: | =8.85°10^(-12) | =C6+1 | =D6+$B$6 | =$B$2°D7 | =$B$7°$B$5°$B$1^2/($B$3 D7)^2 | =E7-F7 |
| 8 |  |  | =C7+1 | =D7+$B$6 | =$B$2°D8 | =$B$7°$B$5°$B$1^2/($B$3 D8)^2 | =E8-F8 |
| 9 |  |  | =C8+1 | =D8+$B$6 | =$B$2°D9 | =$B$7°$B$5°$B$1^2/($B$3 D9)^2 | =E9-F9 |
| 10 |  |  | =C9+1 | =D9+$B$6 | =$B$2°D10 | =$B$7°$B$5°$B$1^2/($B$3 D10)^2 | =E10-F10 |
| 11 |  |  | =C10+1 | =D10+$B$6 | =$B$2°D11 | =$B$7°$B$5°$B$1^2/($B$3 D11)^2 | =E11-F11 |
| 12 |  |  | =C11+1 | =D11+$B$6 | =$B$2°D12 | =$B$7°$B$5°$B$1^2/($B$3 D12)^2 | =E12-F12 |
| 13 |  |  | =C12+1 | =D12+$B$6 | =$B$2°D13 | =$B$7°$B$5°$B$1^2/($B$3 D13)^2 | =E13-F13 |
| 14 |  |  | =C13+1 | =D13+$B$6 | =$B$2°D14 | =$B$7°$B$5°$B$1^2/($B$3 D14)^2 | =E14-F14 |

**TABLE 4-9**  Output of the Spreadsheet of Table 4-8. Solution lies between *y*-values (5) and (6)

|  | A | B | C | D | E | F | G | H |
|---|---|---|---|---|---|---|---|---|
| 1 | VOLTAGE: | 35 | VAL.NO. | COLUMN 1 | COLUMN 2 | COLUMN 3 | COLUMN 4 |  |
| 2 | K: | 60 |  | Y | FM | FE | DIFFERENCE |  |
| 3 | gap: | 5.00E-06 | 1 | 0 | 0.00E+01 | 2.71E-05 | -2.71E-05 |  |
| 4 | side: | 2.50E-04 | 2 | 2.50E-07 | 1.50E-05 | 3.00E-05 | -1.50E-05 |  |
| 5 | area: | 6.25E-08 | 3 | **5.00E-07** | **3.00E-05** | **3.35E-05** | **-3.46E-06** | <--SOLUTION |
| 6 | dy: | 2.50E-07 | 4 | 7.50E-07 | 4.50E-05 | 3.75E-05 | 7.49E-06 | <--RANGE |
| 7 | eo: | 8.85E-12 | 5 | 1.00E-06 | 6.00E-05 | 4.23E-05 | 1.77E-05 |  |
| 8 |  |  | 6 | 1.25E-06 | 7.50E-05 | 4.82E-05 | 2.68E-05 |  |
| 9 |  |  | 7 | 1.50E-06 | 9.00E-05 | 5.53E-05 | 3.47E-05 |  |
| 10 |  |  | 8 | 1.75E-06 | 1.05E-04 | 6.41E-05 | 4.09E-05 |  |
| 11 |  |  | 9 | 2.00E-06 | 1.20E-04 | 7.53E-05 | 4.47E-05 |  |
| 12 |  |  | 10 | 2.25E-06 | 1.35E-04 | 8.96E-05 | 4.54E-05 |  |
| 13 |  |  | 11 | 2.50E-06 | 1.50E-04 | 1.08E-04 | 4.16E-05 |  |
| 14 |  |  | 12 | 2.75E-06 | 1.65E-04 | 1.34E-04 | 3.12E-05 |  |

**TABLE 4-10** Expansion of Spreadsheet of Table 4-9 Starting with Value (3) from Table 4-9 and Smaller *dy*

|    | A        | B         | C        | D        | E        | F        | G          | H            |
|----|----------|-----------|----------|----------|----------|----------|------------|--------------|
| 1  | VOLTAGE: | 35        | VAL. NO. | COLUMN 1 | COLUMN 2 | COLUMN 3 | COLUMN 4   |              |
| 2  | k:       | 30        |          | y        | FM       | FE       | DIFFERENCE |              |
| 3  | gap:     | 5.00E-06  | 1        | 5.00E-07 | 3.00E-05 | 3.35E-05 | -3.46E-06  |              |
| 4  | side:    | 2.50E-04  | 2        | 5.10E-07 | 3.06E-05 | 3.36E-05 | -3.01E-06  |              |
| 5  | area:    | 6.25E-08  | 3        | 5.20E-07 | 3.12E-05 | 3.38E-05 | -2.56E-06  |              |
| 6  | dy:      | 1.00E-08  | 4        | 5.30E-07 | 3.18E-05 | 3.39E-05 | -2.11E-06  |              |
| 7  | eo:      | 8.85E-12  | 5        | 5.40E-07 | 3.24E-05 | 3.41E-05 | -1.66E-06  |              |
| 8  |          |           | 6        | 5.50E-07 | 3.30E-05 | 3.42E-05 | -1.22E-06  |              |
| 9  |          |           | 7        | 5.60E-07 | 3.36E-05 | 3.44E-05 | -7.71E-07  |              |
| 10 |          |           | 8        | **5.70E-07** | **3.42E-05** | **3.45E-05** | **-3.26E-07** | **<--SOLUTION** |
| 11 |          |           | 9        | 5.80E-07 | 3.48E-05 | 3.47E-05 | 1.17E-07   | **<--RANGE**  |
| 12 |          |           | 10       | 5.90E-07 | 3.54E-05 | 3.48E-05 | 5.60E-07   |              |
| 13 |          |           | 11       | 6.00E-07 | 3.60E-05 | 3.50E-05 | 1.00E-06   |              |
| 14 |          |           | 12       | 6.10E-07 | 3.66E-05 | 3.52E-05 | 1.44E-06   |              |

## 4.8 SOLID MODELING AND COMPUTER-AIDED DRAFTING

Simulation, estimation, brainstorming, and other creative processes play important roles in engineering design. In cases where the ultimate goal is a real, physical product, formal sketches and drawings of the object also must be produced. These documents form a key link between design engineers and technicians, fabricators, marketing specialists, customers, and other individuals. Pictorial documentation may appear in one of several forms at various stages of the design process. These forms include informal sketches, isometric views, orthographic projections, exploded views, and solid models. Each of these graphic formats serves a particular design need.

Various methods for generating engineering drawings or *graphics*, have evolved over the years. Before the days of computers, the skill of manual *drafting* (sometimes called "technical drawing") was taught to engineers and technicians in all disciplines. Courses on drafting were common in high school and college curricula, and no self-respecting engineering student would be without a complete set of drafting tools. A typical engineer's drafting kit would have included T-squares, triangular rules, mechanical pencils, erasers, inking pens, and drawing templates. Drafting skills learned in school carried over to the workplace, as the practice of manual drafting was a mainstream activity in most engineering companies. In any given engineering firm, entire rooms would be filled with drafting tables and engineers at work.

Nowadays, computers have virtually eliminated the need for manual drafting skills. Much as the word processor has replaced the manual typewriter, so have numerous *computer aided design* (CAD) software tools replaced the need for engineers with manual drafting skills. Popular CAD software packages, including ProEngineer™ and SolidWorks™, are used by engineering companies everywhere. In this section, we review the key steps involved in using these types of CAD tools for generating engineering drawings.

### 4.8.1    Why an Engineering Drawing?

Consider the engineering drawing of the chassis plate shown below. This object might be used, for example, as part of a battery-powered, scale-model vehicle. Contrast the detailed engineering drawing of the figure with the following written description:

> "The plate should be made from 0.4-mm thick aluminum stock and should be a rectangle 25-mm long by 20-mm wide. It should be drilled with four holes. The first should be located 2.0 mm from the right-hand, 20-mm edge of the plate and 2.5-mm from the upper 25-mm side. The second hole should be located 1.9-mm to the left of the first. These dimensions should be held to a tolerance of 0.1 mm. Both holes should be drilled to a diameter of 0.2 mm with a tolerance of 0.001 mm. These holes should be duplicated using the same dimensions at the other corner of the plate, but located 2.8 mm from the right-hand, 15-mm edge of the plate, and 2.5-mm from the lower 25-mm side."

For most people, the diagram conveys the information much more succinctly than does the written version. The human brain is an extremely efficient image processor, and drawings will almost always surpass written prose an a means for conveying information. The superiority of human imaging power motivates the well-known saying, "A picture is worth a thousand words."

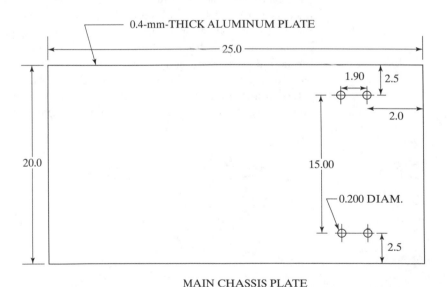

MAIN CHASSIS PLATE

| TOLERANCE TABLE | |
| --- | --- |
| All dimensions in cm | |
| X | ±0.5 |
| X.X | ±0.1 |
| X.XX | ±0.05 |
| X.XXX | ±0.001 |

### 4.8.2 Types of Drawings

As noted previously, engineering drawings come in several widely accepted forms. Categories include hand sketches, isometric projections, orthographic projections, exploded views, and solid models. Each type of drawing has its own particular use in the engineering design process. During the idea-generation phases of a project, *hand sketches* are extremely useful. By quickly drawing things on paper, an engineer can rapidly convey a design concept to other team members. The very act of producing a hand sketch can act as a catalyst for ideas. Hand sketches also are the medium of choice for entering ideas into engineering logbooks.

When a commitment has been made to pursue a particular design concept, more formal types of drawings are in order. The drawing of Figure 4.24 shows the *isometric view* of a simple part. An isometric view is a three-dimensional rendition in which the parallel sides of the actual object are drawn as parallel lines on the page. Isometric views differ from *perspective drawings,* commonly found in classical art and in advertising graphics, in which parallel lines point to a distant "vanishing point." An isometric projection becomes a slightly distorted rendition of the object, but if the part is small and distances are short, the differences will be minor. The isometric view of Figure 4.24, for example, differs little from its equivalent perspective drawing, whereas the isometric and perspective views of, say, a long narrow box would differ significantly. The principle advantage of the isometric view is that it is much easier to draw than a perspective drawing. Also, compared with its related counterpart, the orthographic projection, the isometric drawing provides a "birds eye view" of the object that conveys many of its features at a single glance.

The *orthographic projection* of an object consists of set of two-dimensional projections of the object's front, side, and end views. In some cases, a fourth view may be necessary. Figure 4.25 shows an orthographic projection of the same part described by the isometric view of Figure 4.24. Orthographic projections are principally used by machinists for whom such drawings provide all the information needed to fabricate the actual part. For example, dimensions, tolerances, and machining details are very easy to convey on an orthographic projection. Compared with other types of drawings, orthographic projections are exceptionally easy to draw but require more interpretation on the part of the person trying to read the drawing.

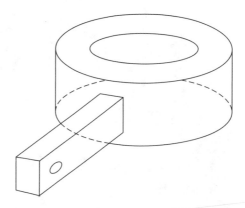

**Figure 4.24.** Isometric view of cylindrical collar with rectangular tab and pin hole.

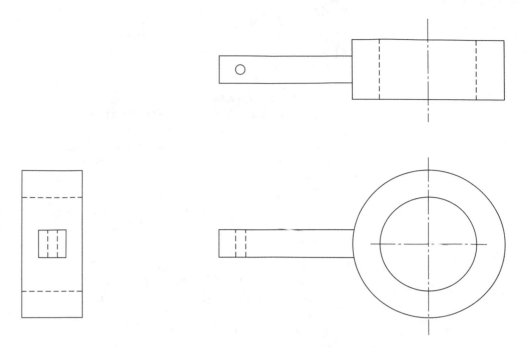

***Figure 4.25.*** Orthographic projection of the part of Figure 4.24.

An *exploded view*, or *assembly drawing*, is used to describe the way in which multiple parts are to be put together to form the working whole. Dotted lines often are used to convey a path to connection or attachment. The diagram in Figure 4.26, for example, shows how the part of Figure 4.25 is to be assembled with other related parts. Although exploded views sometimes can be difficult to draw, they are very useful for conveying information about complex structures.

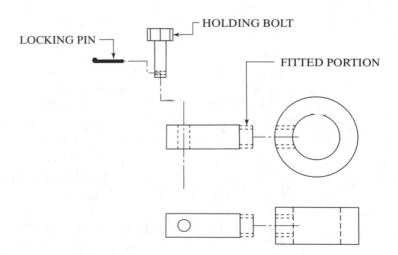

***Figure 4.26.*** Exploded view of several parts shows the way in which they are to be assembled.

The most computationally sophisticated type of drawing produced by a CAD system is called a *solid model*. Unlike isometric and orthographic projections, which depict just the surfaces of an object, a solid model rendition includes information about the surfaces *and* the interior details of the object. A solid model is much more than a simple visualization. It contains a complete mathematical description of the object's material properties as well as its interior and exterior dimensions. This additional information makes the solid model useful for many applications besides viewing. For example, the solid model can be used to predict the object's deformation under applied force, or *stress*, the object's reaction to temperature changes, and its interaction with other parts in the system. At the core of solid modeling lies a computational method known as *finite element analysis* (FEA) in which an object is represented by a large number of interconnected cells, or "elements." A finite element analysis keeps track of the mutual interaction between each cell and its neighbors and computes the behavior of each cell in response to internal and external stimuli. The popular CAD tools ProEngineer and SolidWorks, for example, incorporate finite-element analyses into the solid models of parts and objects.

When an object has been rendered in a CAD tool as a solid model, the latter becomes invaluable when the part is ready for manufacturing. Sophisticated software linked to computer-guided machining tools—lathes and milling machines, for example—can be instructed to fabricate the part directly from its solid model representation. The language used by this class of machines is called *computer numeric control*, or CNC. The CNC system enables an engineer to design a part on the computer screen, then send its CNC code directly to a computer-controlled machine for fabrication from metal, plastic, or other machinable materials. Another method for fabrication directly from solid models is called *rapid prototyping*. In this technique, a rendition of the part suitable for prototype needs is produced using a laser beam that shapes the part from very thin cross sections of plastic resins or paper. The prototype part is assembled, literally layer by layer, by stacking the cross sections.

---

**EXAMPLE 4.10: PRODUCING A SIMPLE PART**

Suppose that you wished to machine a prototype version of the part shown in Figures 4.25 and 4.26. In this example, we illustrate the steps involved in producing the solid model of the part that you could send to a machinist for fabrication. (Alternatively, perhaps you might make the part yourself if you have been properly trained in the use of machine tools.) The steps for producing the solid model drawing are summarized here in a generic way, but they are similar to those one would follow when using specific CAD tools such as ProEngineer and SolidWorks.

*Step 1*. *Open up a new drawing screen:* Open a *new part* screen in the CAD software by choosing **NEW** from the software's **FILE** pull-down menu. A blank screen appears on which the part description will be drawn.

*Step 2*. *Sketch the principal cross section (Figure 4.27):* Using the mouse and keyboard cursors, sketch the part's basic cross section on the screen. Even though the part may have features in all three-dimensions, only its principal cross section (e.g., its top view) need be drawn at this stage. The other views of the part will be generated automatically in a later step. This initial cross section of the part can include some of its machined features. For example, the part in Figure 4.25 includes a drilled hole in its center whose location and dimensions are specified in the initial cross section. The radii of corners and bends in the cross section also can be specified at this stage. This step is sometimes desirable because no machining process can produce perfect angles from solid materials.

*Step 3*. *Dimension the sketch (Figure 4.28):* Major features of the part—lines, circles, and arcs, for example—are selected on the screen and their dimensions are specified in the units chosen for the drawing (e.g., mm, cm, or inches) This step also provides a reference scale for all the other lines and curves that will make up the finished drawing.

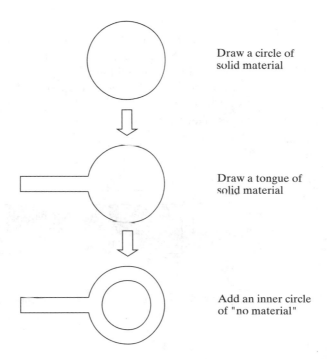

Draw a circle of solid material

Draw a tongue of solid material

Add an inner circle of "no material"

**Figure 4.27.** The principal cross section of the part of Figure 4.25 is drawn on the computer screen.

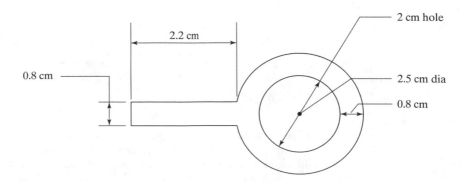

**Figure 4.28.** Dimensions are added to the cross-sectional sketch.

<u>Step 4</u>. *Extrude the cross section (Figure 4.29):* The defined and dimensioned cross section from the previous step is extruded, or "stretched," in the direction perpendicular to the drawing plane to form a three-dimensional version of the part. The extrusion of a circle produces a cylinder, while the extrusion of a rectangle produces a rectangular solid. The extrusion operation leading to Figure 4.29 thus results in a cylindrical solid with a hole in its center plus a rectangular appendage called a tab.

<u>Step 5</u>. *Add features to the extruded part (Figure 4.30):* Once the cross section has been extruded to form a three-dimensional solid model, the three orthographic projections of the part will be available to the designer. Features that are not part of the principal cross section, including perpendicular holes, material cuts, and rounded corners, are added to the object at this stage using the appropriate orthographic view. In this case, a hole whose axis is parallel to the cross-sectional plane is drawn through the rectangular tab, and the top and bottom faces of the tab are trimmed.

<u>Step 6</u>. *Save the file:* The file is saved for future use, printing, and so forth. The solid model rendition of the part is now complete and can be sent as a drawing to other engineers for review in either hard copy or electronic form. Additionally, it can be sent to a CNC-equipped machine tool or to a rapid prototyping machine for computer-controlled fabrication.

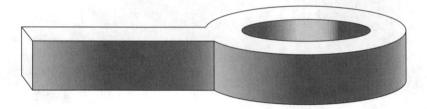

**Figure 4.29.**    The cross section is extruded to form a solid model.

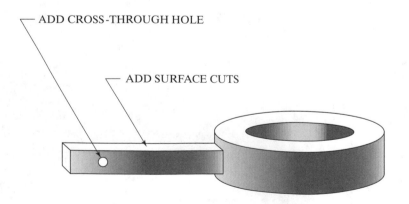

**Figure 4.30.**    Other features are added to the extruded model that were not created during the first extrusion. In this case, a hole whose axis is parallel to the cross-sectional plane is drawn through the rectangular tab, and the top and bottom faces of the tab are trimmed.

## 4.9 SYSTEM SIMULATION

Sometimes, engineers must model the *dynamic* behavior of a system, that is, the way in which the various parts of a system interact and evolve with time. A dynamic analysis differs from the static analyses provided by simulation tools such as ProEngineer and SolidWorks in that the various parts of the system are represented mathematically in block diagram form, rather than as dimensioned objects having physical properties such as density and elasticity. Dynamic simulation tools are useful for solving systems that obey *differential equations.*

One popular dynamic system simulation tool that runs as an appendage to MATLAB is called *Simulink®*. The Simulink user draws a block diagram on the computer that represents the dynamic system to be simulated. The program then determines the relevant equations and sends them to MATLAB for solution and display. This layering of software shells, wherein one program produces code that can be solved by another, is common in software engineering.

**EXAMPLE 4.11: THERMOSTAT CONTROL**

The concept for the following example is derived from an example found in the instruction manual that comes with the student version of Simulink.[2] The block diagrams provided here are more generic than those actually used within Simulink but serve to illustrate the concepts involved. Suppose that you were given the task of designing a temperature control system for a small building heated by a furnace. Because the building has thermal memory, that is, it retains heat for some time when the furnace goes off and requires time to heat up when the furnace is turned on, the furnace-building combination constitutes a dynamic system. The variables of the system include the *desired* temperature (the setting of the thermostat) and the *actual* indoor temperature. The program determines the difference between the two temperatures and turns on the switch if $T_{actual} < T_{thermostat}$. Turning on the switch causes the furnace to produce heat that acts to increase the indoor temperature, but with a time delay $\tau_1$ (sometimes called a *time constant*.) In the meantime, regardless of the status of the furnace, heat continually flows out of the building in proportion to the difference between the indoor and outdoor temperatures. The time constant governing this outward heat flow is $\tau_2$.

A block diagram description of the system is shown in Figure 4.31. The output of the system, produced by Simulink for the parameters $T_{thermostat} = 68°F$, $T_{outdoor} = 32°F$, $\tau_1 = 4$ min, and $\tau_2 = 12$ min, is shown in Figure 4.32. This plot indicates that the temperature falls slowly until $T_{actual}$ falls below $T_{thermostat}$. At that point in time, the furnace turns on and raises the temperature until $T_{actual} = T_{thermostat}$. Note that the building temperature continues to rise for some time after the furnace is turned off. This phenomenon occurs because of the non-zero time delays in the system. Although the system recognizes that $T_{actual}$ has reached $T_{thermostat}$, the time delay between the furnace and the building causes additional heat to flow from the former to the latter.

---

[2] *MATLAB Student Version: Learning Simulink 4* © 2001 The Mathworks, pp 2.3 – 2.5.

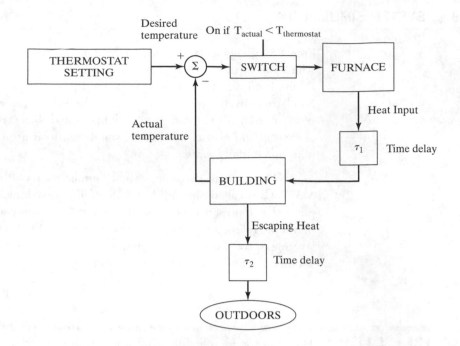

**Figure 4.31.** Block diagram description of the dynamic system of a building and its heating furnace.

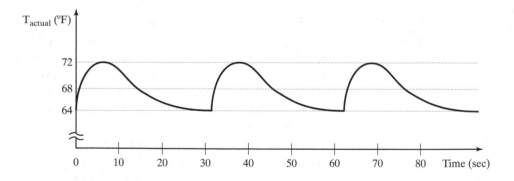

**Figure 4.32.** Results of the simulation with $T_{thermostat} = 68°F$, $T_{outdoor} = 32°F$, $\tau_1 = 4$ min, and $\tau_2 = 12$ min.

## 4.10 ELECTRONIC CIRCUIT SIMULATION

Circuits involving electricity can take many forms. A circuit that delivers energy from batteries, generators, or other sources is commonly called an *electrical* circuit. Examples of electrical circuits include flashlights, the motor drives for electric subway cars, the power lines that deliver electricity to our homes, and the wiring systems inside automobiles and trucks. These systems typically involve only a few basic part types that might include batteries, generators, wires and switches, plus loads that can be modeled as simple resistors. A device such as a resistor is *linear* and obeys Ohm's law: $v = iR$, where $v$ is the voltage applied to the device, $i$ is current that flows as a result, and $R$ is the resistance of the device. The ratio of $v$ to $i$ in a resistive device is constant and is equal to $R$.

In contrast to resistors, semiconductor devices are usually nonlinear. Their voltage-current equations are not simple linear equations but take on more complicated forms. For example, the current-voltage equation for a simple diode is given by the exponential equation

$$i = I_s e^{v/nV_T},\tag{4-24}$$

where $n$, $I_s$, and $V_T$ are constants. The transistor, a three-terminal device, has an even more complicated set of governing equations. Other nonlinear devices include the light-emitting diode (LED), and the integrated circuit. When a circuit contains nonlinear devices, it is usually characterized as an *electronic* circuit, rather than simply as an *electrical* circuit.

Calculating voltages and currents in nonlinear circuits, particularly those with multiple semiconductor devices, can be a daunting task. Even the simple circuit shown in Figure 4.33 cannot be solved by simple algebra. Circuits that contain multiple diodes, transistors, or integrated circuits are readily solved using a circuit simulation tool called SPICE. Several derivatives of this program, including PSPICE, HSPICE, and TSPICE, all use the original SPICE calculation program as a core software engine.

SPICE can model voltage sources (e.g., batteries, the output of *dc* power cubes, and *ac* transformers); resistors; other linear circuit elements called *capacitors* and *inductors*; and all sorts of semiconductors, including diodes, transistors, and operational amplifiers. It also enables the entry of devices with user-specified characteristics. SPICE can perform steady-steady analyses, simulate the evolution of circuit behavior over time, and model the effects of temperature, thermal noise, and random component variations. Output is provided in several forms, including text-based listings of the circuit's various voltage, current, and power dissipation values, as well as plots of selected circuit variables as functions of time. Some versions of SPICE rely on text-based entry of the circuit description. This entry mode stems from the origins of SPICE, which was developed at the University of California at Berkeley in the 1970s as public-domain software for implementation on mainframe computers with punched cards and line printer output. More modern versions of SPICE, such as PSPICE-A/D, provide a graphical interface, a drawing palette on which to enter the circuit description as a schematic, and a list of pre-described commercial parts from which to build the circuit. These latter versions of the program still rely on the original text-based program for core computational operations, and they ultimately translate circuit information entered graphically into traditional text-based SPICE code. Student evaluation versions of SPICE are available for download on the Internet (see, for example, *www.orcad.com/Product/Simulation/Pspice*).

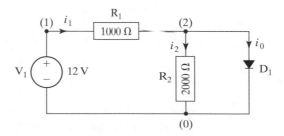

**Figure 4.33.** Circuit to be simulated using PSPICE. Each node of the circuit has been assigned a unique number. The common "ground" node at the bottom is assigned the number 0.

In text-based versions of SPICE, the circuit is described by a set of program statements stored in a file named FILENAME.CIR, where FILENAME is chosen by the user. This data file is retrieved by SPICE during program execution. Each line of the input file contains one program statement only. Each program statement either describes a single part or instructs SPICE with a command. Statements can be continued onto additional lines by entering a "+" sign in the first column of the additional lines. Comment lines are preceded by an asterisk (*) or a semicolon (;).

The typical input file consists of a required *title line*, a set of *element statements* that describe the circuit, and a set of *control statements* that instruct SPICE during program execution. Control statements always begin with a period. The entire file is terminated by the control statement **.END**. In order to simulate a circuit in SPICE, its various nodes must be assigned numbers. One node (usually the most common node, or "ground") must be assigned the number zero.

## EXAMPLE 4.12: NONLINEAR CIRCUIT SIMULATION

The circuit of Figure 4.33 is easily solved using SPICE. A program listing that instructs SPICE to calculate the voltage across, current through, and power dissipated in each circuit component follows. Each node has been numbered, and the bottom, common node has been assigned the number 0.

```
SPICE SIMULATION OF CIRCUIT OF Figure 4.33
*Specify the elements in the circuit.
V1 1 0 dc 12V;12-V dc voltage source between nodes 1
;and 0 (+ side of source at node 1)
R1 1 2 10000; A 10-kΩ resistor between nodes 1 and 2
R2 2 0 10000; A 10-kΩ resistor between nodes 2 and 0
D1 2 0 diode; A diode pointing from node 2 to node 0
.MODEL diode D(Is=1e-5m n=2);A statement that identifies the
;parameters of the diode
.OP;Calculate the operating point (voltage
;and current) of every device
.END
```

Here is a portion of the text output that results from running this simulation:

```
********SMALL SIGNAL BIAS SOLUTION   TEMPERATURE = 27.000 DEG C


    NODE    VOLTAGE   NODE    VOLTAGE
   (   1)   12.0000   (   2)   0.5995


VOLTAGE SOURCE CURRENTS
   NAME           CURRENT
   V1-1.140    E-03


TOTAL POWER DISSIPATION   1.37E-02 WATTS
JOB CONCLUDED
TOTAL JOB TIME  .65
***************************************************************
```

This output indicates that the diode voltage will be equal to about 0.6 V, and its current will be equal to about 1.1 mA. The entire circuit will dissipate 13.7 mW (i.e., convert 13.7 milliwatts of electrical power to heat).

## 4.11    GRAPHICAL PROGRAMMING

A category of software tool called *graphical programming* is extremely useful for certain design tasks. Marketed under commercial software packages that include LabVIEW™, Simulink™, and HP-VEE™, graphical programming languages enable engineers to create programs simply by connecting together visual objects on the computer screen. An object can be a formula, a data source, a display, or a logical function. Objects are tied together with flow arrows much like the boxes of a flowchart to create computer programs. Some graphical programming languages enable the user to interface directly with benchtop instruments, such as multimeters, function generators, and oscilloscopes. Interconnections are made using digital control links, such as the IEEE-488 bus, GPIB (general purpose instrument bus), and HP-IB (Hewlett-Packard instrument bus). A special plug-in bus card is installed in the computer, and cables are fed to each of the benchtop instruments. Data can be sent to and from the instruments via graphical program objects. Connections also can be made to analog-to-digital (A/D) and digital-to-analog (D/A) plug-in cards directly from the computer bus. This environment is ideal for creating automated industrial systems consisting of analog-to-digital and digital-to-analog conversion boards plus a graphical programming interface run from a desktop PC.

An example of a typical graphical program is shown in Figure 4.34. This program is designed to measure the voltage across and current through a motor being tested under load. The mechanical load, expressed in terms of the number of weights applied to a frictional brake, is entered into the program by the user.

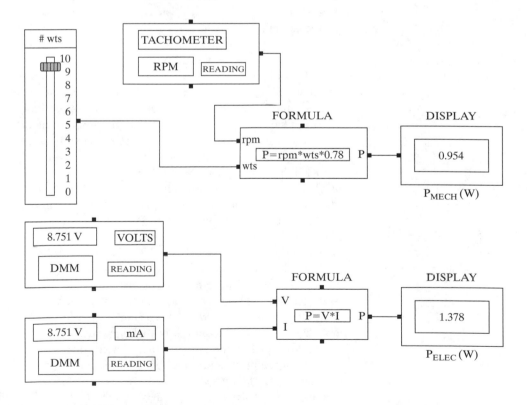

**Figure 4.34.**    Graphical interface program designed to measure the voltage across and current through a motor under test. The mechanical load, expressed in terms of the number of weights, is entered into the program by the user.

## 4.12 MICROPROCESSORS: THE "OTHER" COMPUTER

We usually associate the term "computer" with the desktop PC or networked workstation. In truth, these integrated computational machines represent but a small fraction of all the computers in use today. The most prolific computer, the *microprocessor,* far exceeds in numbers the desktop PC. A microprocessor is a single-chip computer that performs digital logic functions at the fundamental level. Microprocessors are excellent choices for solving many design problems, especially those involving real-time control by computer. These applications are sometimes referred to as *embedded computing*.

In its most basic form, a microprocessor is a computer with no disk drive, keyboard, monitor, or external memory chips, but simply consists of a silicon microcircuit housed in a plastic or ceramic package. A microprocessor lies at the heart of just about every appliance or piece of equipment that requires intelligent control. Microprocessors can be found inside automobiles, microwave ovens, washing machines, children's toys, cellular telephones, fax machines, printers, and, of course, personal computers. The Pentium® chip made by Intel and the Athlon™ chip made by AMD are examples of high-performance microprocessors that are connected to peripheral devices to create desktop PCs. The more numerous microprocessors found inside appliances and machines are usually much simpler devices than Pentium or Athlon chips, but their operating principles are the same. Microprocessors operate in a base 2 arithmetic system of logic rules known as *Boolean algebra*. The rules of Boolean algebra allow a microprocessor to make decisions based on the status of stored or incoming digital data. The collection of these elementary operations comprises a machine-level computer language called *assembly code*. The typical language of assembly code includes such low-level instructions as *read port, store in memory, add, subtract, shift left, shift right*, and *compare byte*. Assembly language also provides simple branching and testing instructions, (e.g., "if ... then" or "branch if equal" commands). More complex operations, such as floating-point calculations, data manipulation, data communication, and control, are performed by piecing together, or *assembling*, sets of the more basic instructions. A microprocessor executes its machine-level instructions extremely rapidly and is ideal for real-time applications.

Simple programs can be created quickly by writing code directly in assembly language. More complex programs typically are written in higher level programming languages such as C, C++, Pascal, or even BASIC. The Basic Stamp™ (*www.parallaxinc.com*) is a ready-to-program microprocessor that is popular with both hobbyists and serious designers. A special PC-based program called a *compiler* reads the high-level program and generates appropriate assembly-language code for the microprocessor. The machine-level assembly code instructions are then transferred, or *downloaded,* to the microprocessor and stored in its permanent, read-only memory (ROM). In the design stage, when a program is being developed, a special, more expensive version of the microprocessor chip is used that permits the code to be erased and rewritten numerous times. The erase procedure may require that the microprocessor be exposed to ultraviolet light for several minutes. The chip package contains a special transparent window for this purpose. Other types of microprocessors use an electrically-erasable memory called EEPROM or sometimes "flash" memory. When the final design of the microprocessor-controlled device is mass produced, less expensive versions of the microprocessor that allow only a one-time burn-in of the assembly code into ROM are used.

Microprocessors communicate with the outside world via sets of terminals, called *ports,* that carry digital signals. A digital signal is a voltage that is either low (0 V), representing the binary number **0**, or high (3 V or 5 V, depending on the system), representing the binary number **1**. A single port consists of a group of 8, 16, 32, or even 64 terminals activated by the microprocessor. A port is bidirectional—it can bring digital data into the microprocessor or send digital data out from the microprocessor. The job of the microprocessor is to perform basic arithmetic or logical operations on the data before sending the results to the outside world. Many microprocessors also include an internal *timer* that allows the device to keep track of time. Others may contain circuits that convert incoming analog data into digital form.

### 4.12.1 Analog-to-Digital and Digital-to-Analog Conversion

Sometimes, microprocessors communicate with the outside world using circuits called *analog-to-digital* (A/D) and *digital-to-analog* (D/A) converters. Most physical quantities, such as temperature, pressure, velocity, and position, are analog quantities. An analog variable has meaning over the entire span of its available range. In electronic systems, analog quantities are typically represented by voltage signals that vary between two limiting values, such as 0 V and 5 V. Sensors have the job of converting a real physical variable, such as position or temperature, into its equivalent analog voltage representation. Suppose, for example, that a sensor produces an analog voltage that represents the position of a sliding bracket. In this sensor, a value of 5 V corresponds to the right-most excursion at 1 cm, and a value of 0 V corresponds to the left-most excursion at −1 cm. If the bracket lies at a position 0.5 cm from the right, for example, the analog sensor monitoring its position will produce a voltage of 3.75 V, or three-fourths of the voltage range between 0 V and 5 V.

In contrast to an analog voltage, a digital voltage, or *bit*, can take on only two states that have meaning. The *on* state of a digital signal corresponds to the binary number **1**, and the *off* state corresponds to the binary number **0**. A digital electronic system usually adopts a voltage level near 0 V to represent binary **0** and a voltage level near either 5 V or 3 V, depending on the system, to represent the binary number **1**. Joining together several digital bits produces a multibit binary number. For example, a collection of eight bits can be used to represent any integer between **0000 0000** (zero) and **1111 1111** (255 in base 10).

An A/D converter accepts an analog voltage as input, compares the incoming voltage to a reference voltage, and produces an equivalent multibit binary number that represents the ratio of the input to the reference voltage. For example, suppose that the A/D converter compares its input to a 5-V reference. If its input is equal to 2.5 V and the conversion is based on eight digital bits, then the converter will represent the input by the digital output **1000 0000**, or 128 in base 10. This output represents 128/255ths, or approximately half, of the analog range, indicating that the ratio of the analog input to the reference is 2.5/ 5 = 0.5. An input of 1 V to the same converter will yield an output of **00110011**, or 51 in base 10. This output represents 51/255ths, or approximately 20%, of the analog range, indicating that the ratio of the analog input to the reference is 1/5 = 0.2.

If a microprocessor sends out a digital signal, it can be converted to analog form by a D/A converter. A D/A converter takes a multibit binary signal as its input and produces an analog voltage as its output. This output will be some fraction of the reference voltage that is supplied to the D/A converter, as determined by the degree to which the converter's binary input number approaches the full scale value represended by all **1**'s. When all the binary bits are **0**, for example, the analog output will be zero. When all the binary bits are **1**, the analog output will be equal to the reference voltage. The availability of A/D and D/A converters greatly enhances the utility of a microprocessor and allows it to communicate with the real physical world. As shown in the next example, microprocessors are used frequently by computer and electrical engineers to help solve design problems.

**PRACTICE!**

1. Convert each of the following binary numbers to decimal (base 10): **0001**, **1000**, **1111**, and **1010**.

2. Write the binary (base 2) presentation of each of the base-10 digits from 0 through 9.

3. An analog-to-digital converter has four bits of resolution, and its reference voltage is 10 V. What voltage increment is represented by each binary digit?

4. A digital-to-analog converter has four bits of resolution, and its reference voltage is 5 V. What output voltage is produced by the inputs **1000** and **1111**?

5. A digital-to-analog converter accepts a 12-bit digital input word. If its reference voltage is 10 V, how much voltage is represented by the least-significant bit? The most-significant bit?

6. How many wires does it take to send a four-bit digital signal from one circuit to another?

7. Draw the serial data stream for the bit sequence **1011**.

**EXAMPLE 4.13: MICRO-PROCESSOR SPEED CONTROL**

This example illustrates the use of a microprocessor for a timing and speed control operation. The application involves the design of a battery-operated car which must traverse a fixed distance within a 15-second time interval. Suppose that a sensor were available that could provide information about the net distance traveled by the vehicle. One form of such a sensor is depicted in Figure 4.35. A transparent disk with eight opaque radial lines passes between the arms of an optical detector. The disk is connected to one wheel of the vehicle. After each 22.5° of rotation of the transparent disk, a line falls between the arms of the optical sensor causing it to send a pulse to the microprocessor. Because the circumference of the wheel is known, the distance traveled by the car between pulses also is known. By simply counting pulses, the microprocessor can compute the total distance traveled by the vehicle, and the microprocessor can issue a "stop" command when it determines that the vehicle has traveled the desired distance. This design approach is preferable to a fixed time interval scheme in that the system will not be sensitive to motor speed, timer errors, or changes in battery voltage.

Using its internal timer, the microprocessor also will be able to determine the car's instantaneous velocity by computing and constantly updating the ratio of distance traveled to elapsed time. Likewise, the microprocessor will be able to set the speed of the car to a value specified by the user at the start of each run. The desired speed could be set via a set of small switches (say, three in number) that are moved to *on* or *off* positions representing logic **1** and **0**, respectively, thereby forming one of $2^3 = 8$ binary numbers lying between **000** and **111**. The microprocessor would first read the switches to determine the desired speed setting, then drive the motor at a level appropriate for that velocity via a digital-to-analog converter. Adopting this clever design would enable the user to choose between a selection of eight preset velocities before each run of the car. A flowchart showing the data input and decision-making requirements of such an embedded microprocessor is shown in Figure 4.36.

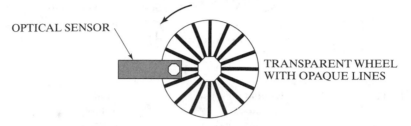

**Figure 4.35.**    Optical encoder wheel produces a digital pulse every 22.5° of rotation.

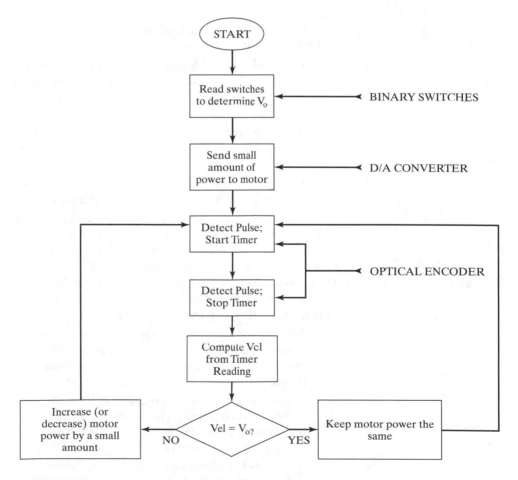

**Figure 4.36.**    Flowchart showing data input and decision making for the microprocessor program.

## PRACTICE!

1. Discuss the way in which a microprocessor might be used to automatically produce timing pulses for a judges starting system.
2. Discuss the way in which a microprocessor might be used to build a digital alarm clock.
3. Identify three appliances or machines that utilize the power of the microprocessor.
4. A cellular telephone contains a microprocessor that provides all operational functions. If you have a cell phone, make a map and/or flow chart of all functions performed by the microprocessor.

## KEYTERMS

| | | |
|---|---|---|
| Estimation | Significant Figures | Dimensions |
| Tolerance | Prototype | Breadboard |
| Reverse Engineering | Internet | Spreadsheet |
| Solid Model | Simulation | Microprocessor |

## PROBLEMS

*Estimation:*   The following three problems relate to a design competition in which a 1-kg battery-powered vehicle must be propelled up a 0.9-m tall ramp within a 15-second time interval.

1. For a run time of 9 seconds, how much power must the battery supply to the vehicle? What would be a reasonable estimate of the average power flow over a 15-second run time?
2. Suppose that batteries are chosen that can supply only 50 mA of peak current. Such a decision might be made to reduce battery weight and produce a lighter vehicle. If vehicle weight is reduced to 0.5 kg, what power flow can be expected? What will be the peak battery current?
3. If motors are chosen with 95 percent efficiency, and mechanical losses are 60 percent, what will be the required battery current?

The following four problems involve vector addition. Vector manipulation is an important skill for estimating forces in mechanical systems. When adding forces or other quantities represented as vectors, the principles of vector addition must be followed. Vectors to be added are first decomposed into their respective $x$, $y$, and $z$ components. These components are added together separately, then recombined to form the total resultant vector. Sometimes, it's convenient to decompose vectors into components lying on axes other than the $x$-, $y$-, and $z$-axes.

4. Two guy wires securing a radio antenna are connected to an eye bolt. One exerts a force of magnitude 3000 N at an angle of 10° to the vertical. The other exerts a force of 2000 N at an angle of 75° to the vertical. Find the magnitude and direction of the total force acting on the eye bolt.
5. A guy wire exerts a force on an eye bolt that is screwed into a wooden roof angled at 30° to the horizontal. The guy wire is inclined at 40° to the horizontal. If the eye bolt is rated at a maximum force of 1000 N perpendicular to the roof, how much tension can safely be applied to the guy wire?
6. A large helium-filled caricature balloon featured in a local parade is steadied by two ropes tied to its midpoint. One rope extending on one side of the balloon is inclined at 20° to the vertical. A second rope located on the other side of the balloon is inclined at 30° to the vertical. If the balloon has a buoyancy of 200 kN, what will be the tension in each of the ropes?

7.  An eye bolt is fixed to a roof that is inclined at 45° relative to the x-axis. The eye bolt holds three guy wires inclined at 45°, 150°, and 195°, respectively, measured clockwise from the x-axis. These wires carry forces of 300 N, 400 N, and 225 N, respectively. What is the magnitude and direction of the total resultant force? What are the components of force measured perpendicular and parallel to the roof line?

    Problems 8–25 can help you develop your design-estimation skills. Discuss them with your friends, and see if you arrive at the same approximate answers.

8.  Estimate the amount of paint required to paint a Boeing 747™ airplane.

9.  Estimate the cost of allowing a gasoline-powered car to idle for 10 minutes.

10. Estimate the daily consumption of electrical energy by your dormitory, residence, apartment building, or home. (Check your estimate against real electric bills if any are available.)

11. Estimate the cost of leaving your computer running 24 hours per day.

12. Estimate the cost savings of installing storm windows on an average-sized four-unit apartment building.

13. Estimate the gross weight of a fully loaded 18-wheel tractor trailer.

14. Estimate the number of single-family houses in your home state.

15. Estimate the number of bolts required to assemble the Golden Gate Bridge.

16. Estimate the number of bricks in an average-sized house chimney.

17. Compute the surface area of all the windows in your dorm, apartment building, or house where you live.

18. Estimate the amount of carpet that it would take to cover the playing field at Chicago's Wrigley Field.

19. Estimate the total mass of air that passes through your lungs each day.

20. Estimate the time required for a stone to fall from sea level to the bottom of the lowest point in the Earth's oceans.

21. Estimate the cost of running a medium-sized refrigerator for a year.

22. Estimate the weight of a layer of shingles needed to cover a single-family, ranch-style house that has a flat roof. Revise your calculations for a pitched roof.

23. Estimate the physical length of a standard 120-minute VHS video cassette tape.

24. Estimate the number of microscopic pits on an average-sized audio compact disk (CD) or digital video disk (DVD).

25. Estimate the number of books checked out of your school library each week.

*Significant Figures:*

26. When calculations are performed, the answer will only be as accurate as the weakest link in the chain. An answer should be expressed with the same number of significant figures as the least accurate factor in the computation. Express the result of each of the following computations with an appropriate number of significant figures:

$$V = (12.9 \text{ mA})(1500\,\Omega)$$
$$F = 2.69 \text{ kg} \times 9.8 \text{ m/s}^2$$
$$F = -3.41 \text{ N/mm} \times 6.34 \text{ mm}$$
$$i = \frac{(1.29 \text{ mA})}{(100)}$$
$$Q = (6.891 \times 10^{-12} \text{ F})(2.34 \times 10^1 \text{ V})$$

27. Evaluate each of the following numerical computations, expressing the result with an appropriate number of significant figures:

$$F = 1,221 \text{ kg} \times 0.098 \text{ m/s}^2$$
$$V = 56 \text{ A} \times 1,200 \text{ ohms} (\Omega)$$
$$x = 76.8 \text{ m/s} \times 1.000 \text{ s}$$
$$m = 56.1 \text{ lb} + 45 \text{ lb} + 98.2 \text{ lb}$$
$$i = \frac{91.4 \text{ V}}{1.0 \Omega}$$
$$P = \frac{(5.1 \text{ V})^2}{1.0 \Omega}$$

## Dimensioning:

28. Measure the dimensions of an ordinary 3.5" computer diskette. Prepare a dimensioned sketch of the diskette, complete with a tolerance table.

29. Prepare a dimensioned sketch of a common coffee cup.

30. Measure the dimensions of a bicycle and prepare a dimensioned sketch of it.

31. Make a dimensioned sketch of this textbook.

## Prototyping:

32. Suppose that 100 mA of steady current flows from a 9-V battery via a timer circuit to a motor. If the controller circuit is 92 percent efficient and the motor 95 percent efficient, how much mechanical power is transferred to the motor wheels (neglecting bearing friction)?

33. Ohm's law states that the voltage across a resistor is equal to the current flowing through it times the resistor value ($V = IR$). Calculate the current flowing through each of the following resistors if each has a measured voltage of 24 V across it: 1 $\Omega$, 330 $\Omega$, 1 k$\Omega$, 560 k$\Omega$, 1.2 M$\Omega$ (Note: 1 k$\Omega$ = $10^3$ $\Omega$; 1 M$\Omega$ = $10^6$ $\Omega$).

34. Ohm's law states that the current flowing through a resistor is equal to the voltage across it divided by the resistor value ($I = V/R$). Calculate the voltage across each of the following resistors if each has a measured current of 10 mA: 1.2 k$\Omega$, 4.7 k$\Omega$, 9.1 k$\Omega$, 560 k$\Omega$, 1.2 M$\Omega$ (Note: 1 mA = $10^{-3}$ A, 1 k$\Omega$ = $10^3$ $\Omega$ and 1 M$\Omega$ = $10^6$ $\Omega$).

35. Kirchhoff's current law states that the algebraic sum of currents flowing into a common connection, or *node*, must sum to zero. Suppose that currents of 1.2 A, –5.4 A, and 3.0 A flow on wires that enter a four-wire node. What current must flow *out* of the fourth node?

36. Kirchhoff's voltage law states that the sum of voltages around a closed path must sum to zero. Three resistors are connected in series to a 9-V battery. The measured voltages across two of the resistors are 5 V and 2.5 V, respectively.

    - What is the voltage across the third resistor?
    - The first two resistors have values of 100 $\Omega$ and 50 $\Omega$, respectively. What is the current flowing through the third resistor?

37. High-power devices, such as thyristors and power transistors, often are mounted on metal *heat sinks*. A heat sink enhances the overall thermal contact between the device package and the surrounding air, leading to a cooler device and a larger power-dissipation capability. Heat removal is important, because excess heat can

cause a catastrophic rise in device temperature and permanent failure. Every heat sink has a heat-transfer coefficient, or *thermal resistance*, $\Theta$ (capital Greek theta), which describes the flow of heat from the hotter sink to the cooler ambient air. The ambient air is assumed to remain at constant temperature. This thermal flow can be described by the equation $P_{therm} = (T_{sink} - T_{air})/\Theta$, where $P_{therm}$ is the thermal power flow out of the device, $T_{sink}$ is the temperature of the heat sink, and $T_{air}$ the temperature of the air.

- A power device is mounted on a heat sink for which $\Theta = 4.5\ °C/W$. A total of 10 W is dissipated in the device. What is the device temperature if the ambient air temperature is 25°C?

- A device rated at 200°C maximum operating temperature is mounted on a heat sink. If the ambient air is 25°C and 25 W of power must be dissipated in the device, what is the largest thermal coefficient $\Theta$ that the heat sink can have?

38.  A switch is a mechanical device that allows the user to convert its two electrical terminals from an open circuit (no connection) to a short circuit (perfect connection) by moving a lever or sliding arm. A *switch pole* refers to a set of contacts that can be closed or opened by the mechanical action of the switch. A *single-pole, double-throw* (SPDT) switch has three terminals: a center terminal that functions as the common lead, and two outer terminals that are alternately connected to the center terminal as the position of the switch lever is changed. When one of the outer terminals is connected to the center terminal, the remaining outer terminal is disconnected from the center terminal.

- Consider the problem of wiring the light in the stairway of a two-story house. Ideally, the occupants should be able to turn the light on or off using one of two switches. One switch is located at the top of the stairs, and the other is located at the bottom. Toggling either switch lever should make the light change state. Draw the diagram of a circuit that illustrates the stairway lighting system.

- Now consider the problem of a *three*-story house in which the lights in the stairwell are to be turned on or off by moving the lever of any one of three switches (one located on each floor). Design an appropriate switching network using two single-pole switches and one double-pole switch. (A DPDT switch has six terminals and can be thought of as two SPDT switches in tandem, with both levers engaged simultaneously.)

39.  A dc motor consists of a multipole electromagnet coil, called the *armature,* or sometimes the *rotor,* that spins inside a constant magnetic field called the *stator* field. In the small dc motors typically found in model electric cars and toys, permanent magnets are used to create the stator magnetic field. In larger, industrial-type motors, such as an automobile starter or windshield-wiper motors, the stator field is produced by a second coil winding.

Current is sent through the rotating armature coil by way of a set of contact pads and stationary brushes called the *commutator.* Each set of commutator pads on the rotor connects to a different portion of the armature coil winding. As the rotor rotates, brush contact is made to different pairs of commutator pads so that the portion of the armature coil receiving current from the brushes is constantly changed. In this way, the magnetic field produced by the rotating armature coil remains stationary and is always at right angles to the stationary stator field. The north and south poles of these fields constantly seek each other, and because they are always kept at right angles by the action of the commutator, the armature experiences a perpetual torque (rotational force). The strength of the force is proportional to the value of armature current, hence the speed of the motor under constant mechanical load is also proportional to armature current.

- Obtain a small dc motor from a hobby or electronic parts store. Connect two D-cell batteries in series with the motor without regard to polarity. Observe the direction of rotation, and then reverse the polarity of the battery connections. Observe the results.

- As an engineer, you are likely to encounter situations in which the rotational direction of a dc motor must be changed by a switch control. Using a double-pole, double-throw switch like the one described in the previous problem, design a circuit that can reverse the direction of the motor using a single switch.

### Reverse Engineering:

40.   Take apart a retractable ballpoint pen (the kind that has a push button on top to extend and retract the writing tip). Draw a sketch of its internal mechanism, and write a short description of how the pen works.

41.   Take apart a common 3.5" floppy computer diskette. Make a sketch of its inner construction, and write a short summary of its various components.

42.   Take apart a standard desktop telephone. Use your investigative methods to develop a block diagram of how the phone works and connects to the outside world.

43.   Suppose that you have been given the assignment to design a desktop stapler. Dissect an existing model from a competitor, draw a sketch of its mechanical construction, and create a parts list for the stapler.

44.   Take apart a common flashlight, draw a sketch of its mechanical construction, and create a parts list from which you could reproduce another.

45.   Imagine that you have discovered an errant, unoccupied space vehicle in an outlying field. Write a report in which you reverse engineer the spacecraft to discover elements of its technology. Examine the vehicle's propulsion system, telemetry, and sensor systems.

### The Computer as an Analysis Tool:

46.   Write a computer program in the language of your choice to calculate the trajectory of a pebble that falls from an airplane traveling at 200 kph. Ignore the effects of air resistance.

47.   Consider a snapping mousetrap bale. Write a computer program in the language of your choice to plot its angle $\theta$ as a function of time from $t = 0$ and $\theta = \pi$ until the time when the bale makes contact with the base at $\theta = 0$. Assume that the bale has a moment of inertia 0.01 kg-m and that the spring exerts a torque of value $0.5\theta/\pi$ N-m, where $\theta$ is in radians.

48.   A capacitor is a device that stores electrical energy. The degree to which a capacitor is charged at any given moment is indicated by how much voltage appears across its terminals. If a resistor is connected across a charged capacitor, then the current $v$ flowing out of the capacitor and into the resistor will be given by the equation $i = v/R$, where $v$ is the capacitor's voltage and $R$ the value of the resistor. The capacitor will respond to the flow of current by reducing its voltage according the equation, $dv/dt = i/C$. Write a computer program in the language of your choice to plot $v(t)$ for the case $v(t=0) = 10$ V, $R = 10$ k$\Omega$, and $C = 100$ $\mu$F. (Note: 10 k$\Omega \equiv 10,000$ ohms; 100 $\mu$F $\equiv 10^{-4}$ farads.)

49.   Draw a flowchart for a computer program that could be used to control the traffic at a busy intersection where two streets cross. Traffic should be allowed to flow

over the east–west route, unless a car stops at the north- or south-bound streets entering the intersection.

Write this program in the language of your choice. Include an input mechanism for indicating the number of cars at each sector of the intersection.

50. Draw a flowchart for a computer program that could be used to control the traffic at a busy intersection where two streets cross. Traffic should be allowed to flow over the east–west route until three cars are stopped at the north street entering the intersection, but only if no car is stopped at the south entrance. If more than three cars become stopped along the east–west route, it should be open to traffic flow, regardless of the number of cars stopped at the north–south streets.

Write this program in the language of your choice. Include an input mechanism for indicating the number of cars at each sector of the intersection.

51. Draw the flowchart for a computer program that can serve as a three-digit password decoder for an alarm system. Each of the digits (0–9) entered into the alarm should be represented in binary form. Choose the last three digits of your birthday year as the password.

Write this program in the language of your choice. Include an input mechanism for each of the entered digits.

52. Draw the flowchart for a computer program that can tally the voting of a 10-person city council. The output should indicate the majority of "aye" or "nay" votes with a logic high (**1**) or logic low (**0**) output. Include a provision for a tie vote.

Write this program in the language of your choice. Include an input mechanism for each of the 10 city council votes.

53. Draw the flowchart and block diagram of a sensor system that can turn on a garden watering system if the temperature rises above 30°C, the sun is not shining on the plants, and the time is not before noon of the same day.

54. Draw the flowchart for a computer program that can be used by a scientific investigator to assess the probability of various events. The system should accept five input signals and provide an output that corresponds to the status of the majority of inputs.

55. Draw the flowchart of a microprocessor program that can be used to sound an alarm in a four-passenger automobile if the ignition is energized but the driver has not put on a safety belt. The alarm also should sound if a passenger is seated but has not put on a safety belt.

56. This problem illustrates the concept of *amplitude modulation.* Suppose that the function $c(t)$ is equal to a triangular waveform that has peak values of +1 and −1 and frequency $f_1$. Similarly, $m(t)$ is a square wave function that varies between +1 and −1 and has frequency $f_2$. Write a computer program that plots the amplitude-modulated waveform described by the equation $v(t) = c(t)[1 + am(t)]$ for the case $f_1 = 10$ Hz, $f_2 = 1$ Hz, and $a = 0.4$. The factor $a$ is called the *modulation index,* and hertz (Hz) has the units of cycles per second.

57. Draw a flowchart that will implement the system described in the following memo:

To: Xebec Design Team
From: Harry Vigil, Project Manager
Subject: ATM Simulator

Our client for this project is in charge of teaching programs for mentally challenged students in a regional school system. The curriculum followed by these students includes practicing such common tasks as calling on the telephone, going to the supermarket, making change, and taking the bus. Their teacher has requested that we develop a machine that can simulate the functions of an automatic teller machine (ATM) such as one might find at any local bank. The availability of such a system will allow the teacher to train students without tying up an actual bank ATM machine.

Your task is to design and build such a simulator using the components and materials of your choice. The details of the simulator's operation should be fine-tuned after your initial meeting with the customer; however, here is a list of specifications that you can use as a guide in preparing your initial project proposal and technical plan:

- The simulator should be self-contained with no actual contact with the outside world.

- It should realistically simulate such features as user inquiry, prompting for password, type of transaction, and dollar amounts.

- The simulator should include its own set of entry keys or buttons, display device, printer, and dispenser for (simulated) money.

- The simulator should be triggered into operation by the insertion of an ATM-type bank card. The decoding and interpretation of information stored on the inserted cards is not a necessary feature of the system. An acceptable solution could instead involve storing passwords (possibly as an updatable list) inside the simulator; this list would be activated by the insertion of any card of the appropriate size.

*Spreadsheets:*

58. Write a spreadsheet program that computes the trajectory of rubber-band launched projectile described by Eqs. (4-11)–(4-18). Your spreadsheet calculations should have cells in which you can enter key parameters of the problem such as dimensions, launch angle, and amount of rubber band stretch. Your spreadsheet should yield the same results as those found in Example 4.2, as summarized by the plots of Figure 4.5. Use an array of cells to indicate the position of the projectile at various points along its trajectory.

59. Suppose that you are trying to come up with a budget for an engineering design project. Your salary is about $4,000 per month, and the technician with whom you work earns about $2,500 per month. You need $5,000 for materials and supplies, $1,200 for traveling to a sales meeting, and you have to contribute 8 percent of the direct dollars you spend to support clerical staff. In addition, you have to pay 80 percent of the entire contract, including the 8 percent clerical charges, to overhead that supports the general operation of the company. Write a spreadsheet program that can help determine the maximum number of person-months that you'll be able to charge to the project if the total budget must not exceed $100,000. You'd like your technician to work at least half time on the project.

60. Write a spreadsheet program that will help you determine the center of gravity of all the passengers on a small commuter airplane. The airplane has ten rows of four seats each, with two seats on either side of the aisle, for a total of forty seats. The centers of the aisle seats are located 0.5 m from the aircraft centerline, and the centers of the window seats are located 1.2 m from the aircraft centerline. The rows are spaced 1 m apart. Your program should compute the center of gravity measured relative to the first row of seats ($y$-coordinate). Assume that you know

the weight of each passenger. Feel free to do your analysis in either kilograms or pounds.

61. A film-manufacturing plant produces standard 35-mm photographic film for cameras. The film is produced on large rolls between 500 m and 2,000 m long. Each large roll might have a width between 0.5 m and 2.0 m in steps of 0.5 m. That is, there are four possible values for the width. The large rolls are sliced into 35-mm strips and packaged into consumer-sized canisters for 12, 24, or 36 pictures per strip. Assume that each picture requires 35 mm of strip length, and that the 35-mm wide strip inside each canister must allocate 10 percent of its length to a trailer and leader (i.e., unexposed film at the start and end of each roll). Write a spreadsheet program that will allow you to compute the total number of film canisters of each size obtainable from a large roll for various percentage allocations of the three values of shots per canister. Your spreadsheet should have the following user entries: width of large roll, length of large roll, percent each of 12-, 24-, and 36-shot canisters desired from the entire roll. Assume that the slicing process generates no wasted film.

62. Suppose that the owner of a ferry boat has asked you to design a system for helping to load cars and freight in the most balanced way. Write a spreadsheet program that will help you determine the moment of inertia about the center of gravity of the ferry due to all the passengers and freight on the ferry. The ferry is to have 40 parking spaces, each 2.7 m × 4 m, and it should accommodate as many 2.7 m × 8 m shipping containers as possible. The total cargo area for the ferry is 50 m × 100 m. Assume a weight of 1,000 kg per vehicle and 6,000 kg per shipping container.

## Real-Time Computer Control:

63. Convert each of the following binary numbers to decimal (base 10):

| | | |
|---|---|---|
| a. 11 0000 | d. 1010 1010 | g. 1111 1111 0000 |
| b. 10 0110 | e. 1110 0111 | h. 1010 0101 1010 |
| c. 01 1110 | f. 1100 1011 | i. 0111 0111 0111 |

64. Write the binary (base-two) presentation of each of the following numbers:

| | | |
|---|---|---|
| a. 8 | d. 256 | g. 5,061 |
| b. 127 | e. 1,001 | h. 6,977 |
| c. 255 | f. 2,087 | i. 10,657 |

65. How many binary bits does it take to represent an analog voltage to a resolution of at least 1 mV if the range of the analog voltage is 0–5 V?

66. An analog-to-digital converter has eight bits of resolution, and its reference voltage is 10 V. How large an increase in measured signal voltage is required to increment the value of the binary representation of the signal by one bit?

67. List at least 10 possible sources of electrical noise that might effect a sensitive electronic circuit. Each item on your list should be something likely to be encountered in everyday life (e.g., *not* an alien spacecraft).

68. How many wires does it take to send a 12-bit digital signal from one circuit to another?

69. Computers communicate with other computers and peripherals using either serial or parallel data links. When a parallel link is used, the connection to the computer consists of one wire for each of the bits in the digital word, a common return ground wire, plus additional wires for sending synchronizing signals. The latter

are needed so that the receiving device will know when to read each digital word sent by the transmitting device.

When a serial link is used, data bits are sent one at a time. In a *synchronous link* system, one wire is used by the transmitting device to send the data bits in sequence, one wire is used for return ground, and a third wire is used to send a synchronizing signal. The latter is used by the receiving device to determine the timing between each data bit. In an *asynchronous link* system, typical of the type used to communicate over telephone modems, long-range networks, and the Internet, only one wire pair is available for signal transmission. Data-bit synchronization requires that the sending device and receiving device both be set to the same BAUD (for *bits audio*), or bit timing rate. Such timing is never perfect, however; if left uncorrected, the BAUD timing of the receiving device will drift apart from the BAUD timing of the sending device. To ensure that the bit-timing sequences will match, the receiving device resets its timer after each digital word. It knows when the received word ends, because the transmitting word appends a *stop-bit sequence* to each one. After the stop bits are sent, the transmitting device sets the data line to the value **1** as a prelude to sending its next word. It also adds a *start bit* of value **0** to the beginning of each word so that its arrival will be unambiguous.

- A particular computer sends data in eight-bit packets, or *bytes*. Determine the content of the two bytes represented by the data sequence shown in Figure 4.37. The start bit, from which you can determine the time interval per bit, precedes a second **1** bit. The stop-bit sequence consists of two data bits held high.

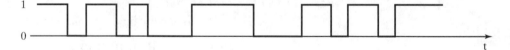

**Figure 4.37.** Asynchronous serial data stream.

- Draw the serial data stream for the byte sequence (**1001 1100**) (**0001 1111**) (**1010 1010**).

70. Indicate the type of data link (parallel, synchronous serial, or asynchronous serial) used by a PC to communicate with each of these peripherals:

- Parallel printer interface
- External 56 kbps modem
- MacIntosh printer interface
- Flatbed scanner
- Electronic piano (MIDI interface)

71. A *pulse width modulation* motor drive system applies voltage to a motor while adjusting the *duty cycle,* or time interval, over which full voltage is applied. The current that flows when voltage is applied will be determined by the motor speed and internal resistance. The average power consumed by the motor will be equal to the time average of the voltage–current product. The pulse–width modulated waveform is produced in response to a digital data signal from a computer module. Write a program in C, MATLAB, or the language of your choice that can produce the required voltage waveform. Your program should accept a binary or

decimal number between 0 and 255 and then produce an output that is high (logic **1**) for an amount of time proportional to the input value.

### Analog-to-Digital and Digital-to-Analog Conversion

72.  Analog-to-digital interfacing is an important part of many computer-controlled engineering systems. Although most physical measurement and control involves analog variables, most data collection, information transmission, and data analyses are performed digitally. A/D and D/A circuits provide the interfaces between analog and digital worlds.

   A D/A converter produces a single analog output signal, usually a voltage, from a multibit digital input. One common conversion algorithm produces an analog output proportional to a fixed reference voltage as determined by the equation

$$v_{OUT} = \frac{n\,V_{REF}}{(2^N - 1)}$$

where $N$ is the number of bits in the digital input word, $n$ is the decimal value of the binary number represented by all the input bits that are set to **1** in the digital input word, and $V_{REF}$ is a reference voltage. When $n$ is equal to $(2^N - 1)$, $v_{OUT}$ is equal to $V_{REF}$.

- Suppose that the input to an 8-bit D/A converter is **0010 1111** with $V_{REF} = 5$ V. Find the resulting value of $v_{OUT}$.
- A 10–bit D/A converter is fed the input word **00 1001 0001** and is given a reference voltage of 5 V. What is the output of the converter?
- What is the smallest *increment* of analog output voltage that can be produced by a 12-bit D/A converter with a reference voltage of 10 V if the algorithm previously shown is used?
- What is the largest analog output that can be produced by an 8-bit D/A converter if $V_{REF} = 12$ V?

73.  An A/D converter compares its analog input voltage to a fixed reference voltage and then provides a digital output word $B$ given by

$$B = \frac{int\ v_{IN}\,(2^N - 1)}{V_{REF}}$$

where the operator *int* means "round to the nearest integer." This encoding operation is called *binary-weighted* encoding. A full-scale binary output (all bits set to **1**) occurs when $v_{IN} = V_{REF}$.

- An 8-bit binary-weighted A/D converter has a reference voltage of 5 V. Find the analog input corresponding to the binary outputs **1111 1110** and **0001 0000**.
- Find the binary output if $v_{IN} = 1.1$ V.
- Find the resolution of the converter.
- Find the additional voltage that must be added to a 1-V analog input if the digital output is to be incremented by one bit.

74.  Boolean algebra is a system of logic used by many computers. In Boolean algebra, variables take on one of two values only: TRUE (logic **1**) or FALSE (logic **0**). Boolean operators include AND, OR, and NOT. The AND operator is represented by a dot between variables, (e.g., $Y = A \cdot B \cdot C$ means that $Y$ is true if $A$, $B$,

and $C$ are *all* true). The OR operator is represented by plus signs (e.g., $Y = A + B + C$ means that $Y$ is true if one or more of $A$, $B$, or $C$ is true). The NOT operator, represented by an overbar, simply reverses the state of the variable (e.g., $\overline{A} = 0$ if $A = 1$).

- Verify the following equations in Boolean algebra:

$$A \cdot B + A \cdot \overline{B} = A$$
$$(A + B) \cdot (B + C) = B + A \cdot C$$
$$(A + B) \cdot (\overline{A} + C) = \overline{A} + \overline{C + B}$$

- DeMorgan's theorem states that the Boolean expression $\overline{A \cdot B \cdot C}$ is equivalent to $\overline{A} + \overline{B} + \overline{C}$. Similarly, $\overline{A + B + C}$ is equivalent to $\overline{A} \cdot \overline{B} \cdot \overline{C}$. Verify both forms of DeMorgan's laws.

### Use of Computers:

75. Discuss the way in which a computer or microprocessor (single-chip computer) might improve your approach to designing the following products:

    - an all-electronic telephone answering machine
    - a tachometer and an odometer for a bicycle
    - an energy-saving light switch
    - a smart clothes iron that shuts off after 1 hour without use
    - a data logger for measuring part weight in a quality-control system
    - a voice-synthesized device for a speech-impaired person
    - a digital alarm clock
    - a system for producing Gantt charts for homework assignments

76. Identify a system or entity in your school, home, or place of work that would benefit greatly from the introduction of a computer system. Write a short summary explaining why.

77. Identify a system or entity in your school, home or place of work that suffers because one or more computers were introduced inappropriately. Write a short summary explaining why.

78. Take a survey of a select group of people who use computers. (The people you choose could be those who live on your dorm floor, attend one of your classes or are in your extended family, for example.) Make a list of how many hours per day each person uses a computer and the approximate percentage of time spent at various tasks on the computer. Examples of tasks include word processing, spreadsheet use, CAD drafting, computation, e-mail, etc.

79. Make a list of at least 10 appliances or machines that you encounter on a regular basis, exclusive of desktop computers, that utilize the power of a microprocessor.

80. Make a list of at least five appliances or machines that you encounter on a regular basis whose function could be greatly improved by the use of a microprocessor.

81. Write the flowchart of a microprocessor program needed to run a desktop telephone with the following features: Tone dial, redial button, flash button, memory (8 registers), and hold.

# 5

# The Human–Machine Interface

Have you ever noticed how some machines are easier to use than others? Do you find that some products appeal to your sense of touch and sight, while others simply perform their designated function? Do some software programs seem extremely user friendly, while others seem impossible to operate? Easy-to-use items have one thing in common: an excellent human–machine interface. The human–machine interface defines the way in which a person interacts with an engineered product. It evokes the senses of touch, sight, and hearing. A good product is easy to use and "feels" right. It becomes an extension of the user's motor and cognitive functions. Its features were not simply included just because they could be. Rather, the product was developed from the start with the needs of the user in mind. The designer of a good product worries about *how* the device will be used as well as *what* the product must do. A good product is one in which the entire function and purpose of the device has affected the design process. Experienced engineers know that the human–machine interface is one of the first things that should be addressed during the initial stages of product development. In this chapter, we examine the role of the human–machine interface in engineering design. Throughout this chapter, the word "machine" is defined to mean any mechanical, electrical, industrial, structural, biomedical, or software entity designed by an engineer.

## SECTIONS

- 5.1   How People Interact with Machines
- 5.2   Ergonomics
- 5.3   Cognition
- 5.4   The Human–Machine Interface: Case Studies

## OBJECTIVES

*In this chapter, you will:*

- Learn how people interact with machines.
- Understand the importance of the human-machine interface.
- Examine case studies of good and bad human-machine interfaces.

## 5.1   HOW PEOPLE INTERACT WITH MACHINES

The pages of engineering history are littered with tales of products that fulfilled a useful function but were hopelessly inept at providing an adequate human–machine interface. One canonical example of this principle is the programmable VCR. In a television commercial that aired in the early 1990s, a satisfied user of an infomercial-marketed device for programming a VCR via voice command proudly proclaims, "Hey, even I can't program my VCR by hand, and *I* have a *Master's* degree!" Indeed, much humor has been derived from the notion that seemingly intelligent people are content to let their VCR clocks blink at 12:00 rather than tackle the intricate maze of its programming functions. The problem with most underutilized VCR features is seldom the intelligence of the user, but rather that most of us simply do not have the time to master hard-to-learn features provided by the manufacturer. Many other simple consumer products suffer from this deficiency. Examples include various versions of the digital watch, the programmable CD player, cellular telephones, PC operating systems, word processors, fax machines, coffee makers, and microwave ovens. Even the latest models of automobiles are laden with features that elude the owner who is usually intent on simply driving the car. Without a simple, easy-to-remember sequence of programming steps, the features of the most intricate and complex machines will lay idle most of the time.

## 5.2   ERGONOMICS

The human–machine interface influences our attitude toward the most mundane of devices. Some doors seem to require less effort to push open. Some knife-and-fork sets are more appealing to use than others. A favorite chair feels more comfortable than all the others. The best of these products were designed with careful consideration of *ergonomics,* the science of how the body interacts with machines. Ergonomics focuses on the size, weight, and placement of objects and control devices that interact with the human body. The average span of the human arm, the swing frequency of an arm, the height of the eyes above a tabletop, the spacing between fingertips—these are all examples of things that must be considered when designing the physical layout of a product. Consider the case of the driver's position in the automobile, a product that has been fine-tuned for nearly 100 years. Auto makers give careful consideration to the placement of the steering wheel, gearshift lever, brake and accelerator pedals, heat and air-conditioning controls, radio knobs, rear view mirrors, and windshield aperture. Cars are designed for the average human body while maximizing as much as possible the range of physical attributes that can be accommodated. By how much should the position and height of a seat be adjustable? Will adding the expense of a tiltable steering wheel gain more market share because drivers on the fringe of the ergonomic range will find the car appealing to drive? This approach to vehicle design has served the driving public for most of the history of the automobile, although occasionally unforeseen problems occur. The concern over airbags that surfaced in the mid 1990s, for example, illustrates a design principle run amok. The airbag is an amazing device that can be credited with saving countless lives. But like all devices in the automobile cockpit, it was designed with the average driver in mind. It inflates at a "chest level" defined for a person of average height. Children and short adults can receive a lethal blow to the head should an airbag inflate unexpectedly. This unforeseen problem led to the revised recommendation that children be confined to the rear seats of all automobiles having passenger-side airbags. While this advice provides a temporary solution, the hazard remains a problem for short adult drivers and front-seat passengers and has led to redesigned air bags of the so-called "stage II" variety.

The study of ergonomics has produced a body of anthropometric data that can be used in designing anything that involves the interaction of a human and a machine. Tables of statistics on arm length, arm span, joint location, limb bending angles, and turning radii can be found in numerous references in the literature, including those listed at the end of this chapter. These data can be used to choose the size of knobs, location of openings, spacing of push buttons, and location of display devices. One goal should be to avoid awkward positions or physical actions on the part of the user. The software developers who wrote many of the early programs that ran under the Microsoft DOS™ operating system, for example, did not have the advantage of today's ubiquitous graphical user interface, mouse, or scrollbars. All computer commands had to be initiated by keyboard sequences. These programmers were keenly aware of the importance of ergonomics, however. They took great advantage of the close proximity of the function keys to the letter keys on the early PC keyboards (see Figure 5.1) to provide easy-to-use keyboard command sequences. Entire command sets—cursor motions, scrolling, cut-and-paste, find-and-replace— could be effortlessly evoked with minimal finger motions by pressing nearby Fn keys or simple CTRL-LETTER combinations. To facilitate these commands, the CTRL key was located to the immediate left of the letter A, where it could be easily pressed by the little finger of the left hand. Conversely, the master "panic" reset sequence was specifically chosen to be the very hard-to-engage CTRL-ALT-DEL triple key combination. By today's standards of graphical user interfaces and the point-and-click mouse, the ergonomically efficient key combinations of the mid 1980s may appear cumbersome, especially because the Fn key sets are now located in an upper row above the numerals 0 to 9 on most keyboards, and the CTRL key has moved to the bottom row of keys. But at the time, these key combinations embodied the essence of good ergonomic design.

### 5.2.1   Putting Ergonomics to Work

The rules of ergonomics are not difficult to learn. Although some of the more esoteric guidelines are the province of experts, most involve simple common sense.

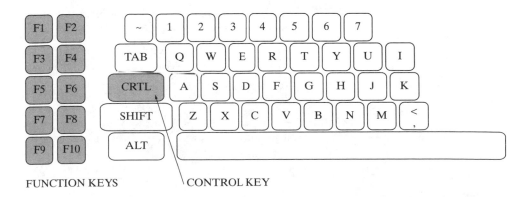

FUNCTION KEYS            CONTROL KEY

**Figure 5.1.**   Left-hand side of the original 88-key computer keyboard. The function keys were located in two rows to the left of the letter keys, and the CONTROL key was located just to the left of the letter A. These key locations represented good ergonomic design.

Button controls should be kept within a finger span unless they are intended for occasional use only. Knobs and valves should be kept within an arm's reach, and those linked to a common function should be grouped together. Visual information should be kept within line of sight. Display devices should be located so as not to require constant turning or bobbing of the head. Command sequences or operations should follow a logical order and should be easy to remember. Some of these design principals are so ingrained in our subconscious that we notice them only when they are violated. If you sat down behind the wheel of a car, for example, and found the ignition key to the left of the steering wheel, you'd note that the car's layout did not feel quite right. If you opened a new software program and found the FILE pull-down menu in the upper right-hand corner of the screen instead of upper left-hand corner, you might think it strange. The principles of ergonomics share a unique symbiosis with the shape of our bodies, the dimensions of our limbs, our shared expectations, and even elements of our culture.

**PRACTICE!**

1. Measure the shoulder-to-finger-tip arm span of your own body. Compare to the height of your desk and distance from its edge to your computer keyboard. Is there a correlation?

2. The standard desk height is 76 cm (30 in). Measure the height of your waist above the floor. Is there any correlation?

3. Measure your own floor-to-shoulder height and compare to the standard 106 cm (42 in) height of an electrical switch above the floor.

4. Measure your own head diameter and compare to the unseparated span of the earphones of a stereo headset. Is there any correlation?

5. Sit in car, and measure the distance between the bottom of your foot and the accelerator pedal as the seat is moved from its extreme forward and backward positions. What range of driver leg length do you think the car can accommodate?

6. Examine the FILE pull-down menus of up to 10 computer programs and write down the order in which the various entries appear. Compare and contrast the differences in a simple table.

7. Examine the minimum and maximum height of buttons located in several elevators. Compare these numbers with the range of human arm span.

8. Measure the height-to-depth ratio of several staircases. Do you observe a pattern?

9. Find the height of the standard entry door. What percentage of the population will not fit under this door opening without stooping?

10. Measure the minimum and maximum base-to-cord lengths of several desktop telephones in the stretched and unstretched positions. Compare these figures to your own arm span.

## 5.3   COGNITION

Nearly every engineered device requires learning on the part of the user before it can be operated. The typical user wants to have control over a product and fully understand its use. *Cognition* refers to the way in which the user learns about the device, masters its features, and becomes familiar with its various characteristics. A well-engineered device provides the user with a short learning curve and a consistent set of rules of operation. As an example of this principle, consider the now-standard graphical user interface (GUI) found in most computer software. Have you noticed how the FILE and EDIT pull-down menus are always located in the upper left-hand corner of the screen? They are placed in this location because the user *expects* to find them there. A user will consider a program easy to use when it builds upon features learned from prior use of similar products.

Another example of this principle can be found in the common automobile. The gearshift, directional signal lever, and horn button are always in the same place. Every car driver learns to operate these controls and expects them to be in the same place in every automobile. One aspect of car operation that has no consistency is the location and operation of the headlight switch. How often have you driven a strange car only to fumble momentarily while trying to figure out how to turn on the headlights?

In designing a product or system, care should taken to make its operation easy to learn. As you design a new device, mimic the operating principles of similar devices. Borrow functional sequences, structural details, or command sequences. Place controls where they are likely to be found on similar machines, or, at the very least, in logical places. We expect a light switch to be located along a room's interior wall, just inside the unhinged edge of the door. This location is logical, given the way one enters a room. Finding a switch placed anywhere else contradicts our learned behavior. The same can be said for the direction of the volume control on a radio or stereo system. We subconsciously expect the volume to be increased by turning the knob clockwise. There is no hard logic to this choice, but it resonates with our learned notion that the rotational direction of an analog clock corresponds to marching forward, or increasing something. Acknowledging the importance of cognition in engineering design requires that we design products and systems whose operation is easy to learn, easy to remember, and consistent in operation with other, similar products.

## 5.4   THE HUMAN–MACHINE INTERFACE: CASE STUDIES

Engineers can learn a great deal about design by studying the successes and failures of other engineers. While grand, large-scale failures, such as falling bridges and collapsing roofs, have gained much notoriety in educational circles, small-scale failures, particularly in the arena of the human–machine interface, are equally worthy of study, because they illustrate the impact of the small design decisions engineers make every day. This section cites several case studies of both well-designed and poorly designed human–machine interfaces.

In the discussions to follow, the happy-face symbol (☺) denotes an example of a good human–machine interface. The sad-face symbol (☹) denotes a bad interface, and the neutral-face symbol (😐) denotes a "mixed bag."

**EXAMPLE 5.1: THE HAND-SET TELEPHONE DIAL** ☹

Until the mid-1970s, most all telephones had rotary dials. The integrated-circuit technology needed to integrate DTMF (dual-tone, multiple-frequency, or Touch-Tone™) capability into telephones was neither economical nor practical. A marvel of electromechanical design, the rotary dial, shown in Figure 5.2, allowed the user to insert a finger in the hole corresponding to the desired digit. Rotating the dial from the chosen digit to the hook and releasing it caused it to send out the proper number of electrical current pulses to the central telephone station. The main problem with the rotary dial was that it was mechanical, and therefore more prone to failure. It also was much slower than its soon-to-appear Touch-Tone replacement.

The Princess™ telephone, first marketed by Western Electric, signified the beginning of a new trend in telephone technology. Its small rotary dial was moved from its customary desktop base to the center of a new, flat-profile headset. This design represented a departure from the traditional desk phone that had been the standard since the 1940s. Its compact and utilitarian arrangement fulfilled the design goal of providing an instrument that did not require the user to return to the base dial to make another call. This arrangement was ideal for someone lounging on a couch and making repeated phone calls. Might there have been a connection between the original name "Princess" and the image of a teenager in the 1960s? The engineers who designed the Princess might not have conceived of the pervasiveness of DTMF dialing that we enjoy today. Indeed, much of our modern world of communication—modems, cellular telephones, cordless phones, voice mail, pagers, telephone banking—all rely on the DTMF system. Nevertheless, the physical form of the Princess phone became the model for many of these later devices. All cordless phones, cellular phones, and many table-top models are designed with the DTMF keypad located in the center of the headset. The keypad-in-headset concept is so pervasive that it has become a de facto standard. This approach to keypad location, however, violates an important principle of ergonomic design. It makes the product much more difficult to use. In the typical scenario involving voice mail or automated telephone functions, the user first listens to an announcement, then presses the appropriate key selection. For most people, the latter operation requires looking at the keypad while pressing the selected key. The user must repeatedly hold the headset

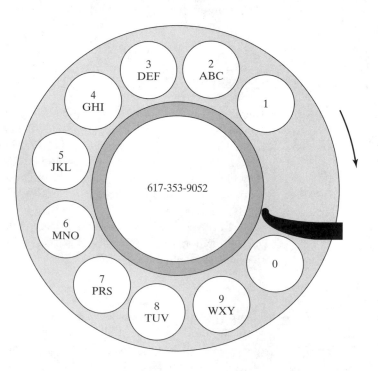

**Figure 5.2.** The rotary telephone dial.

to the ear, remove it from the ear, hold it in front of the face to press keys, and then return the headset to the ear to listen for the results or the next automated instruction. It is much easier to handle automated telephone dialing or voice-mail instructions when the keypad is located in front of the user, as it is in a conventional desktop phone. (The latest technology circumvents the deficiencies of the keypad-in-headset phone by responding to spoken numerals. These systems are slowly gaining popularity and are likely to supplant the deficiencies brought about by the original Princess design.)

**PRACTICE!**

1. Consider two telephones, one with the keypad in the headset, and one with the keypad on its desktop base. Imagine that you are trying to access an automated flight arrival system for a major airline. Estimate the extra time per call required to enter the numbers from a headset keypad.

2. Measure the longest span between two keys on a typical desktop telephone. Measure the index-to-middle finger span on your own hand. Which is larger?

3. Do a survey of 10 or more toll-free customer service call-back lines. Before the main menu is finished, dial "0" for "Operator." Determine the number of sites for which prematurely dialing 0 (versus some other digit) actually brings a live person to the phone.

4. Call the toll-free information line of any major airline. How many keys must you press, in addition to the toll-free number, in order to find out the arrival time of a flight?

5. Compare the locations of the digits 0 to 9 on a telephone keypad with the locations of these digits on a calculator keypad. Is there a correlation?

**EXAMPLE 5.2:**
**THE**
**MOUNTAIN**
**BIKE ☺**

The mountain bike, shown in Figure 5.3, has become the choice of recreational bike riders everywhere. From the casual urban user to the most serious mountain trekker, riders are drawn to the mountain bike's comfort, durability, and ease of use. It's evolution to prominence provides an example of good ergonomic design. During the 1960s, the bicycle of choice by serious riders was a three-speed model that had a single gearshift lever mounted near the right-side grip of the handlebars. This bicycle, of British design and dubbed the "English bike" by Americans, replaced the balloon-tired "coaster brake" models which were soon relegated to kids-only status. A decade later, the 10-speed bicycle dominated the bikeways. With its lightweight frame, large choice of gear ratios, and sleek tires, the 10-speed was built for speed and efficiency. It soon became the favorite of teenagers, college students, and even adult riders. Its design was cloned from the bicycles that had been used for years by professional racers. As illustrated in Figure 5.4, the brake leverswere mounted on the front of curved handlebars, and the gearshift levers were mounted on the foremost strut of the bicycle frame. The placement of these controls required the rider to assume a hunched over, albeit aerodynamically efficient, position. In order to change gears, the rider had to let go of the handlebars with one hand in order the reach the gear levers. Although the hunched-over position is very efficient for racing and extended cross-country riding, most casual users found it cumbersome. Minor solutions, such as placing the gearshift levers on the handlebar post and the addition of brake-lever extensions, became available toward the end of the 10-speed's heyday.

***Figure 5.3.*** A rack of mountain bikes.

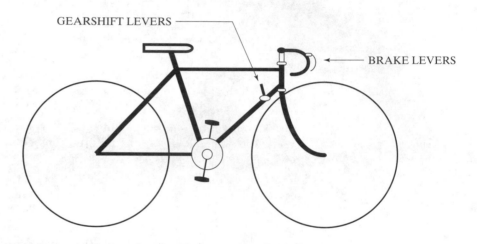

***Figure 5.4.*** The classic 10-speed bicycle.

Sometime in the late 1980s to early 1990s, the mountain bicycle began to appear in bicycle shops. It had a straight handlebar, its brake levers were within easy reach of an upright rider, and its gearshift levers were integrated right into the hand grips. The rider no longer had to let go of the handlebars to shift gears! By the mid 1990s, the mountain bike had become the bike of choice by all but serious cross-country riders and racers. Its strong frame and rugged tires, designed to assault mountain paths, also proved to be worthy of the bumpiest of city streets. The success of the mountain bicycle can be attributed to its designers' careful attention to ergonomics and their willingness to consider the riding functions and needs of the end users.

## EXAMPLE 5.3: THE TOGGLE LIGHT SWITCH ☺

Electricity first came into widespread use at the turn of the 20th century. Early methods of wiring and insulating were primitive, because the plastic materials that prevail today were not readily available. Indeed, most of them had not yet been invented. The first wall switches for lighting were large, rotational devices made from metal and ceramic. The user turned a light on and off by turning a ceramic knob. Sturdy and virtually indestructible, these early pioneers of electrical switching had one major drawback. It was impossible to tell if the switch was on or off simply by looking at it. Later designs added a pointer to the rotary knob, but determining the position of the switch still required a careful examination of the switch. The much improved toggle switch came into use sometime in the late 1930s to early 1940s. Its simple up-down design has persisted to this day. Its cognitive function is second nature to us all: "Up" means "on" and "down" means "off." (The only exception to this rule occurs in the three-way switches used for hallway lighting.) If you've ever encountered a room light switch that is mounted upside down, you've probably experienced a momentary sense that something was wrong as you tried to flip on the light.

The toggle motion of today's light switches provides another important ergonomic benefit. Unlike its rotary predecessor, it can be switched with no hands at all. An elbow, knee, hip, stick, or even well-placed nose can do the job. This feature is helpful to users carrying shopping bags, individuals with special needs, small children, or the adults that carry them.

**EXAMPLE 5.4: THE REAR VIEW MIRROR AND THE SUN VISOR ☹**

The 2001 model of a popular automobile incorporates numerous high-tech features, including steering-wheel mounted AM/FM radio controls, separate driver and passenger-side temperature controls, and an in-vehicle roadside-assistance call system. Despite these useful technological advances, the automobile has a simple but significant design flaw. When the driver attempts to turn the sun visor to the down position, it collides with the rear view mirror (see Figure 5.5). Either the sun visor must remain in only the partial down position, or the rear view mirror must be pushed out of adjustment in order to accommodate the visor. This example underscores the need to test even the simplest ergonomic aspects of any new design.

**Figure 5.5.** The sun visor in this car hits the rearview mirror.

## EXAMPLE 5.5: THE CASSETTE TAPE PLAYER ☹☺

The Sony Corporation makes an aftermarket radio/cassette player for use in automobiles. Because the Sony name is normally associated with quality, most buyers would assume that the unit has been designed sensibly. (An *aftermarket* product for an automobile is one that is installed later by the owner rather than at the factory by the automobile maker. In contrast, an original equipment manufactured product, or *OEM*, is one that may be made by a third-party vendor but is specifically designed to be built into the vehicle at the factory. An OEM product seldom displays the name of the actual manufacturer.)

Although the Sony player functions well *electronically*, it has a serious ergonomic flaw. The slot for cassette tapes is located *above* the digital readout for the clock and radio station selector buttons. The radio will not work with a tape inserted all the way into its slot, as one would do to play a tape. In order to get the radio to work, the user must eject the tape from the cassette slot. If a tape is ejected but left in the tape slot (as one might need to do, for example, in order to momentarily switch to the radio news while listening to a tape), the tape blocks the digital display from the driver's line of sight, as illustrated in Figure 5.6. This misplacement of the cassette slot means that a finished tape has to be removed entirely from the unit in order to view the clock or choose a radio station for listening, adding a needless distraction for the driver. The poor choice of functional layout suggests that the unit was designed by desk-bound engineers who never actually tried out the unit under typical driving conditions.

A better example of tape cassette design can be found in a unit made for OEM installation in Toyota automobiles. As shown in Figure 5.7, this product has four buttons next to the tape cassette slot labeled AM, FM, TAPE, and CD respectively. When a tape is playing, pressing the AM or FM buttons will stop the tape and turn on the radio without ejecting the tape cassette. This feature allows the driver to see the digital readout without having to remove the tape from the unit.

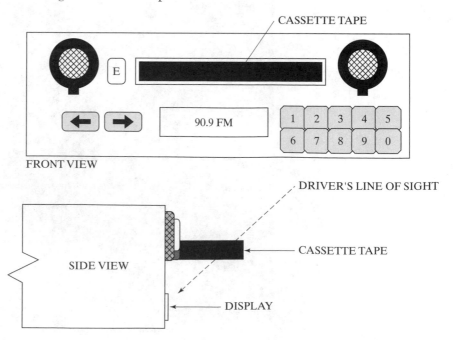

**Figure 5.6.** The misplaced tape cassette slot. A tape cassette left in its slot in the ejected position blocks the digital readout from the driver's line of sight.

**Figure 5.7.**   The tape cassette in this tape player stays inside when the AM or FM buttons are pressed.

**EXAMPLE 5.6: THE BATTERY CHARGER** ☺☺

The battery charger shown in Figure 5.8 is made by Olympus as an accessory to its line of digital cameras. It allows the user to externally charge the four AAA-size nickel-metal hydride (Ni-MH) batteries that power the camera. Unlike many similar products, in which adjacent batteries must be inserted in alternating directions, the four batteries on the Olympus charger are all inserted with the plus (+) side facing in one direction. This example of good cognitive design greatly simplifies the use of the product.

Interestingly, the company has not built this cognitive feature into the camera itself. When inserting freshly-charged batteries into the camera, the user must alternate the direction of the batteries. The diagram etched on the inside of the battery cover flap that indicates proper insertion directions is small and difficult to read.

**EXAMPLE 5.7: THE DIGITAL CLOCK** ☺

Digital clocks can be found everywhere in our society. Display technologies have progressed to the point where economical electronic time pieces are sometimes less expensive than the batteries that power them. The earliest models of digital clock incorporated power-hungry red light-emitting diode (LED) displays into ac-powered desktop alarm clocks. Advances in integrated-circuit technology produced red LED watches that required the user to press a button to see the time. (Red LEDs require so much power that constant illumination via battery power was not feasible.) Soon thereafter, power-miserly liquid crystal displays were developed along with single-chip clock circuits, opening the door to cheap, reliable wrist watches and other timing devices. The attractiveness of electronic time-keeping technology and its robustness has caused digital time to all but replace the mechanical timekeeping devices that served us well for over 500 years.

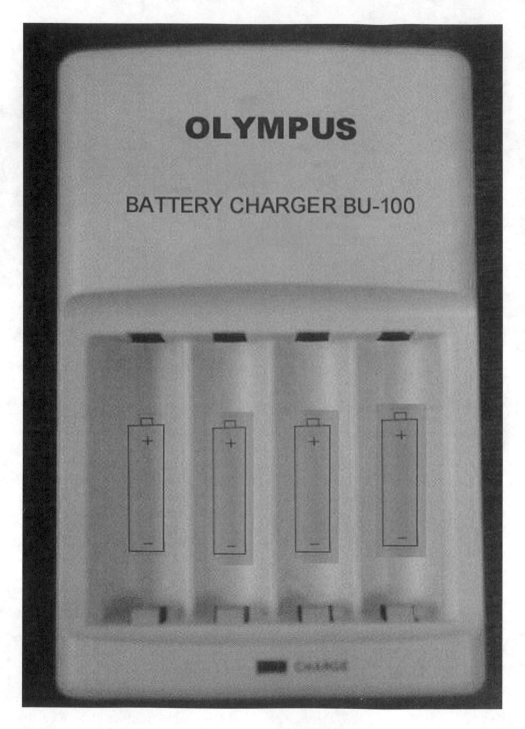

**Figure 5.8.** The batteries in this charger are all inserted in the same direction.

Engineers who study cognitive principles have come to realize that digital clocks have a fundamental disadvantage when compared with their analog counterparts. Although a digital clock allows you to know the exact time instantly, most people are more interested in how much time *remains* before some critical event. Examples include, "How much time is left before my class is over?" or "How many minutes do I have left to get to my dentist appointment?" Viewing time from a digital clock requires that you do the necessary arithmetic in your head. This mental exercise can take a few extra seconds when time is provided to the nearest minute, for example, subtracting 9:48 from 10:00 to determine that you have but 12 minutes left to arrive at your 10 o'clock destination.

While your brain can readily do serial math, it functions much better as an image processor than as a calculator. In this task, the human brain is incredibly fast. No machine has yet been invented that can beat the human brain at general pattern recognition. Viewing the time 9:48 on analog clock hands enables you to estimate and subtract all in one glance as you instantly determine that you have about 10 minutes left to get to your appointment. The contrast between digital and analog clocks provides an example of the often-ignored discord between technical progress and human cognition.

**EXAMPLE 5.8: THE GYMNASIUM LIGHTING SYSTEM ☻**

The town of Brookline, Massachusetts, renovated its high-school gymnasium. The school community was very proud of the upgraded facility, which included dual-use seating so that the room could be used for assemblies as well as for sports functions. As part of an overall energy-saving strategy at the school, the overhead lights in the new gym were equipped with motion detectors. In the absence of motion on the central gym floor, the lights would gradually dim and turn off.

The new gym was a huge success. It became the site of numerous basketball games, physical education classes, and school dances. The first time the room was used for a general assembly, however, a critical design flaw was discovered. A very famous speaker had been invited to address the student body on an important social issue. The speaker's podium was set up at one end of the room, and students and faculty sat on the sides in bleachers and chairs. Ten minutes into his speech, the lights began to dim, eventually leaving the entire assembly in darkness. The motion detectors, sensing no movement in the center of the gym, proceeded to turn off the lights under the mistaken conclusion that the room was empty. By the way, the design engineers had failed to provide a manual override for the motion detectors. The assembly continued in darkness.

**PRACTICE!**

1. Consider the tape cassette shown in Figure 5.6. Draw a redesigned unit that would eliminate the line-of-sight problem described in Example 5.5.

2. Sketch the design of a digital clock that eliminates the cognition problem discussed in Example 5.7.

3. Specify the details of an energy-saving lighting system that would allow the gymnasium described in Example 5.8 to be used for movement-minimal assemblies.

**EXAMPLE 5.9:
THE BEEPING
DIGITAL
THERMO-
METER** ☺

Digital thermometers for measuring human body temperature, like the one in Figure 5.9, arrived on the scene sometime during the 1980s. Originally too expensive for home use, their price has become so low that digital thermometers have largely replaced their older (and more dangerous) mercury and glass counterparts. Some brands include an extremely useful feature that embodies good cognitive design. If the thermometer has not been turned off after being removed from the patient's mouth and set aside, it will begin to beep after a few minutes. Much like the beeping of a phone left off the hook, the beeping thermometer reminds the user to turn it off so that its battery will not discharge. Were this simple feature not incorporated into the device, it might be discharged and useless the next time a family member had a fever in need of monitoring.

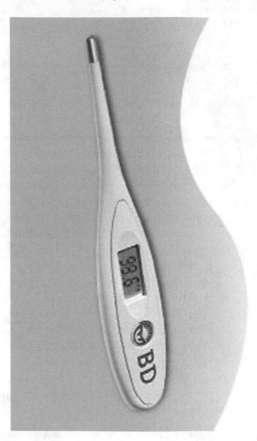

**Figure 5.9.**   Digital thermometer for measuring human body temperature.

**EXAMPLE 5.10:
THE DESIGNER
LAVATORY
FAUCET** ☹

A famous maker of decorative plumbing fixtures has marketed the designer lavatory faucet shown in Figure 5.10. This fixture is a captivating composite of chrome and brass that adds to the décor of any well-appointed bathroom. The straight, cylindrical handles rise cleanly above the curved, inlaid base. This visually appealing faucet has but one problem: When the smooth, cylindrical handles become the slightest bit wet, they are nearly impossible to turn. The faucet handles provide a good example of form taking needless precedence over user function.

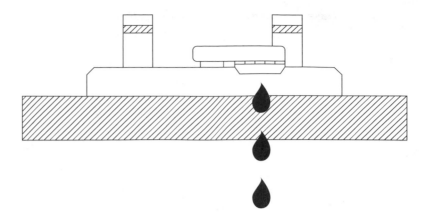

**Figure 5.10.** The designer laboratory faucet.

**EXAMPLE 5.11: THE DIGITAL VOICE RECORDER** ☺

The digital voice recorder shown in Figure 5.11 is an example of good ergonomic design. Its teardrop shape fits right into the palm of the hand, and its button controls are within easy reach of the user's index and middle fingers.

**Figure 5.11.** Digital voice recorder fits perfectly in the palm of the hand. All controls are within easy access of a finger or thumb.

**EXAMPLE 5.12: THE MARINER'S COMPASS ☺**

Good ergonomic design need not be limited to the world of high tech. The simple mariner's compass, used by sailors for centuries, provides an example of bad ergonomic design turned good. The first compasses were made by floating pieces of lodestone, a naturally occurring magnetized rock, in water or other liquids. Free to turn in any direction, the lodestone unfailingly pointed in the north-south direction, providing an important navigational aide to sailors. As technology progressed over the centuries, the hard-to-find lodestone was replaced by magnetized iron and steel, eventually evolving into the floating compass rose design shown in Figure 5.12(a). This basic form of the compass, with its horizontally floating disk and glass bubble top, persists in many ships to this day. The problem with this design is that the disk can be viewed only from the top, because the printed surface of the compass is horizontal. The helmsman must look away from the horizon to glance down at the compass, leading to fatigue on long ocean voyages. Adding a mirror inclined at 45° to allow viewing from a horizontal line of sight does not help, because it reverses the apparent rotational direction of the compass and confuses the helmsman. The solution to this problem is a simple one. A much-improved, redesigned compass is shown in Figure 5.12(b). It replaces the floating disk with a cylindrical shell mounted on a jeweled pivot. The compass markings are printed on the vertical edge of the cylinder, allowing the entire compass to be mounted on a vertical wall of the cockpit. Mounted in a proper location, it can be viewed with only a slight downturn of the helmsman's eyes, rather than a complete lowering of the head. Fatigue during long voyages is greatly reduced.

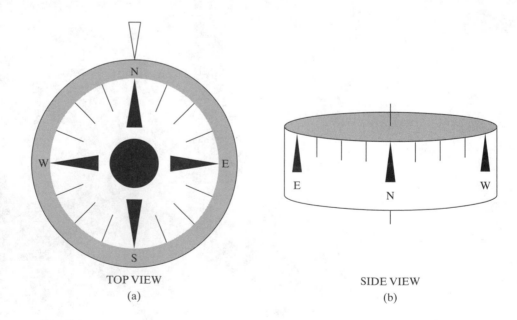

TOP VIEW
(a)

SIDE VIEW
(b)

**Figure 5.12.** The mariner's compass: (a) original horizontal compass design; (b) revised vertical wall compass design.

**EXAMPLE 5.13: THE SQUEEZE KETCHUP CONTAINER** ☺

While not glamorous, the lowly squeeze ketchup container (Figure 5.13) is an example of superb ergonomic design. It's easy on the hand because its compliant walls are just the right thickness to make squeezing effortless. Its translucent walls let the user see just how much ketchup is left inside. Its small pointed spout always dispenses the ketchup exactly where it's needed. Its glass bottle counterpart requires that the user pour ketchup by madly thumping on the base of the overturned bottle. Whoever invented the plastic squeeze ketchup container deserves recognition for a superior design.

**Figure 5.13.** The squeeze ketchup bottle.

**EXAMPLE 5.14: THE CEREAL DISPENSER** ☹

The large container shown in Figure 5.14 was made for storing breakfast cereal and other dry goods. The body of the container is much larger than the span of a typical hand grip, but the item is well designed and includes a recessed edge that is easily gripped for pouring. Its top includes a sealing flap that permits easy opening without requiring removal of the tight-fitting lid. The profile of the container as seen from the sides and top, including the lid, is shown in Figure 5.14.

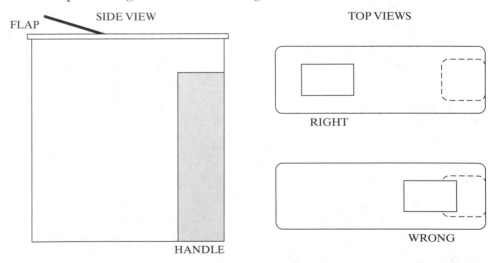

**Figure 5.14.** The cereal dispenser. Top right: Lid flap placed on the correct side, opposite the handle. Bottom right: Lid flap placed incorrectly on the same side as the handle.

The design of this container has one major ergonomic flaw: The user can fill the container and put on the lid in either direction. No mechanism exists for forcing the user to put the opening flap on the side opposite the hand grip. An unfamiliar user may position the lid on the wrong side of the container, leading to an awkward situation at pouring time.

**PRACTICE!**

1. Consider the lavatory faucet shown in Figure 5.10. Draw a revised design that would be easier to use but would retain the modern lines of the unit shown.
2. Consider the cereal dispenser of Figure 5.14. Draw a modification to the product that would ensure that the cover is put on correctly each time.

**EXAMPLE 5.15: THE HOSPITAL REVOLVING DOOR ☺**

A large hospital recently renovated its main lobby. The previous building design had included revolving doors as a way to save energy by limiting air exchange during pedestrian traffic flow. Entering or leaving the hospital by wheelchair required use of large swinging doors located beside the bank of revolving doors. The problem with this arrangement is that many wheelchairs passed through the hospital each day. The hospital has a policy of requiring all released patients to be delivered to their cars by wheelchair. In addition, many incoming patients are wheelchair-bound, and many parents enter the hospital wheeling baby strollers. All these individuals were forced to use the swinging doors, bypassing the energy-saving feature of the revolving door.

The design of Figure 5.15 solved the problem. By designing a revolving-door cavity as an elongated rectangle with semicircular ends, and by designing an articulated door with outer wings, the architects were able to produce a revolving door that accommodates wheelchairs, baby strollers, and regular pedestrian traffic. The new design expedites traffic flow and saves energy at the same time.

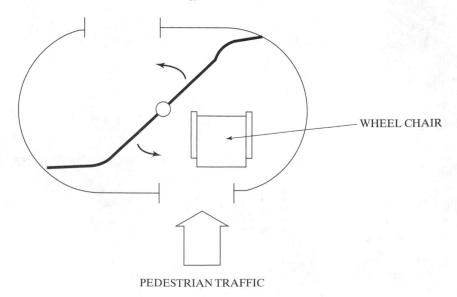

PEDESTRIAN TRAFFIC

***Figure 5.15.*** The hospital revolving door.

**EXAMPLE 5.16: THE PAPER CLIP** ☺

The simple office paper clip is a marvel of good ergonomic design. First patented around 1900 (and attributed, most say, to the Norwegian scientist, mathematician, and engineer Johan Vaaler), the basic paper clip of Figure 5.16 has survived virtually unchanged for over 100 years. The different lengths of its long and short tongues make it extremely easy to clip over several pages using only one hand, and the different widths of its two tongues—one fits inside the other—gives it added clasping strength. Although manufacturers have attempted to come up with numerous improved designs for the paper clip, none has succeeded in displacing the popularity of this indispensable office item.

***Figure 5.16.*** The common paper clip.

**EXAMPLE 5.17: WEB FRAMES— AN ORGANI- ZATIONAL PARADIGM** ☺

The first Internet Web pages were of very simple design. All were written directly in HTML (hyper-text markup language) and may have included a picture or two. Navigating though a lengthy page meant a lot of vertical scrolling and was difficult at best. The onset of the "frames" mode of page design has greatly improved both the ergonomic and cognitive efficacy of Web pages. In the frames mode, information can be retrieved with precision and minimal scrolling. The contrast between single page and Web frame design is analogous to the relationship between music on tape, which must be accessed serially, and music on a CD, which can be accessed randomly. The simple software design change embodied in frames capability has made information on the Web ever more assessible to a broad spectrum of computer users.

**PROFESSIONAL SUCCESS: BECOME AWARE OF THE HUMAN/MACHINE INTERFACE**

Formal study can help you learn about the human/machine interface, but direct observation in your own life is equally valuable. Be observant. If something seems difficult to use, try to figure out why. Think about how you might redesign the product to make it easier to use. Devise, in your own mind, modifications that will improve the product. By becoming aware of the machines and technology around you and identifying the design flaws of others, you'll become more skilled at designing your own human/machine interfaces.

**KEY TERMS**

Ergonomics
Case Studies

Cognition

Human-Machine Interface

**REFERENCES**

W. E. WOODSON, B. TILLMAN, and P. TILLMAN, *Human Factors Design Handbook: Information and Guidelines for the Design of Systems, Facilities, Equipment, and Products for Human Use*, 2d ed. New York: McGraw-Hill, 1992.

K. R. FOWLER, *Electronic Instrument Design: Architecting for the Life Cycle.* New York: Oxford University Press, 1996.

R. W. BAILEY, *Human Performance Engineering: A Guide for System Designers*, Englewood Cliffs, NJ: Prentice-Hall, 1982.

## Problems

1. Compile a list of five objects or devices that illustrate good ergonomic design. For every entry on your list, come up with a counterexample of poor ergonomic design.

2. Prepare a case study of an example of good or improved ergonomic design.

3. Prepare a case study of an example of poor ergonomic design.

4. Perform a telephone survey of 50 adults at random. Of those who own and use a VCR, ask how many regularly make use of its programmable features.

5. Survey people who use computers. Divide the list into those who use a conventional mouse, those who use a trackball, and those who use another pointing device. Of those in the first two categories, determine how many are happy with their pointing device and how many wish there existed something "better."

6. Design an experiment in which you draw a facsimile of the view as seen from the driver's seat of an automobile. Change the location of one feature of the layout (e.g., the location of the ignition key or gear shift lever). Show your drawing to a collection of test subjects. Record the amount of time that it takes each test subject to identify what is out of place.

7. Draw a diagram of a room with doors, windows, and furniture. Place the door handle on the same side as the hinges. Show your diagram to a number of test subjects. Record the amount of time that it takes each test subject to identify what is out of place.

*Ergonomic Measurements:*

8. Measure a typical classroom chair in your school. Record the following dimensions: front-seat lip to floor; front-to-rear length of seat surface; front-seat edge to backrest; rear-seat edge to middle of backrest. Compare these dimensions to their corresponding body measurements on 10 individuals and compile your data.

9. Find light switches in 20 different buildings (not just different rooms in the same building). Measure the height of the switch above the floor, and find the average value and the standard deviation. Now measure the elbow-to-ground height of 30 different people. Also determine the average value and the standard deviation, and compare to the light switch measurements.

10. Measure the shoulder-to-fingertip arm span of 20 adults. Determine the average length, the minimum, the maximum, and the standard deviation. Do 20 people provide a large enough sample? Should you obtain measurements of more people?

11. Measure the floor-to-eye height of 10 seated people who work regularly at a computer. Compare their measured height to that of the center of the monitor screen on their computer. Ask each individual to estimate how long he or she can work at

the computer before needing a break. See if there is any correlation between fatigue and monitor placement.

12. Find 25 or more volunteers who are willing to walk a fixed distance of approximately 30 m (about 100 ft.). Count the total number of paces that each person requires to span the distance. Determine the average value and the standard deviation.

13. Measure the head circumference of 30 or more individuals. Determine the average, minimum, maximum, and most frequent head circumference measured to the nearest half centimeter.

*Histograms:*

The following set of problems involves the use of the *bin histogram.* A histogram is a graphical plot that shows the number of members of an ensemble of data in various categories. For example, the histogram of Figure 5.17 shows the total number of occurrences of each letter of the alphabet in the text of this problem up to and including the end of this sentence. Similarly, the histogram of Figure 5.18 shows the distribution of end-of-semester grades in a particular engineering class. Engineers often use histograms to display data and sort information.

14. Ask 100 full-grown people to tell you their height. Create a bin histogram that shows the distribution of heights in your sample pool.

15. Ask 100 adults to tell you their weight. Create a bin histogram that shows the distribution of weights in your sample pool.

16. Measure the exact height and width of 30 different desks. Plot your data in bin histogram form.

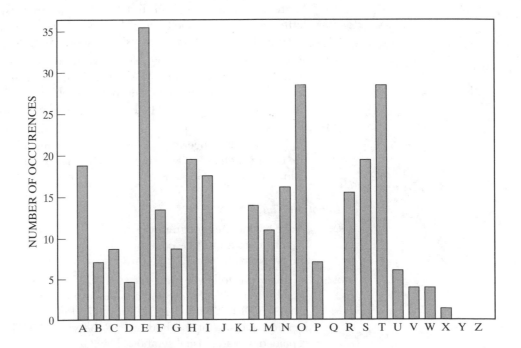

**Figure 5.17.** Bin histogram showing the frequency of the letters of the alphabet appearing in paragraph text.

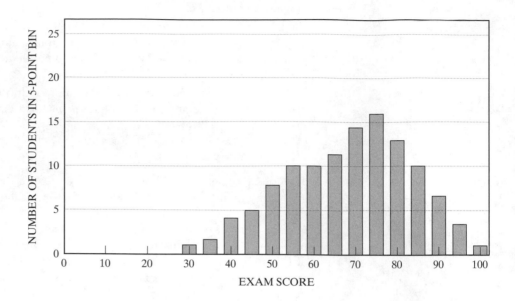

**Figure 5.18.** Bin histogram showing the distribution of grades in the engineering class.

17. Measure the length and width of at least 50 marked parking spaces in your community. Exclude parking spaces for individuals with handicaps. Create two histograms, one for the length and one for the width, that show the distribution of parking space sizes.

18. Write a computer program to determine the keystroke frequency of each of the letters A–Z by someone typing The Gettysburg Address (or any similar document of your choice). Plot your results using a bin histogram.

19. Find 25 people who ride bicycles. Measure the length of their legs from hip joint to the bottom of the foot with shoes on. Now measure the height of their bicycle seats above the ground. Plot a histogram of the ratio of seat height to leg length. Is there any obvious pattern? Is there any gender correlation?

20. The objective of this problem is to determine the most common choices of color for personal passenger cars. Find a location along a highway or busy road. Record the color of at least 100 passing cars. Display your data, from most to least popular color, in bin histogram form.

*Reaction Time:*

21. Write a computer program to track the keystrokes of someone typing a document of your choice into the computer. Find and record the average time that it takes the typist to activate each key after the entry of the preceding letter.

22. Perform the following test on several classmates. Prepare a card of simulated pushbuttons with the printed commands ON, OFF, LEFT, RIGHT, UP, DOWN, TURN LEFT, TURN RIGHT, and STOP. Prepare a second card in which the location and size of the buttons are the same, but the printed words have been replaced with the visual symbols shown in Figure 5.19.

Now devise a set of keystroke sequences that simulate the navigation of a remote-control robot around a fictitious maze. Ask 10 or more friends to press the simulated buttons upon your verbal commands. Keep track of how much time it takes each person to complete the sequence. Which do you think will lead to a faster reaction time: the printed keys or the graphically labeled keys?

23. As you dictate the following passage, compare the time required for each of 20 people to write it on paper in longhand against the time required to type it into a computer: *Six saws saw six cypresses.* (Translate this sentence into French and it comes out sounding like, *"See see see see see prey."*)

24. Write a game on your computer that displays time from a made-up clock. For some participants, the time should be shown in analog form with hands. For others, it should be shown in digital form. As you repeatedly flash times on the screen, ask participants to press a key to stop the flashing when the displayed time is 12 (or whatever) minutes before another time that you've specified. For example, if you specified 5 minutes before 11:20, the correct answer would be 11:15. Have your computer program keep track of how much time elapses between the display of the correct answer and the pressing of the stop button. On average, is there a difference between reaction times to analog versus digital displays?

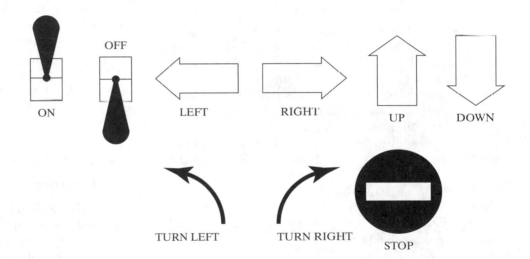

***Figure 5.19.*** Visual commands for up, down, etc.

# 6

# Engineers and the Real World

## 6.1 SOCIETY'S VIEW OF ENGINEERING

Now that you've decided upon engineering as a career and have gotten this far in the book, it's time for the bad news: If you become an engineer, you will be destined for oblivion. Think about it. The press and the broadcast media rarely cover the lives of engineers. The individual exploits of politicians, criminals, actors, financiers, glamour models, and sports figures all receive coverage in the press, but one seldom hears about the personal details of any important engineer. The only time that engineers do receive detailed coverage is when a major failure causes some public catastrophe. No TV show or movie has ever highlighted an engineer as the main character. Rather, engineers are always portrayed as either the brains behind some villain's master plot or the nerdy fellow who provides technical support as the hero saves the day.

In truth, society at large depends on the work of engineers every day. The public expects devices and systems designed by engineers to work flawlessly—all the time—and only tout the faults of engineers when things fail. Most people take for granted the national power grid that magically causes electricity to appear from ubiquitous wall sockets. Only when those sockets go dead during blackouts does the public cry that "something must be done," implying that the blackout has occurred because of the incompetence of the engineers who supply electricity.

What is the reason for this stilted view of engineers? Is it because engineers are robotic people with flat personalities and no imagination? Is it because only high-school nerds become engineers? Is it because engineers are incapable of being leaders? Far from it. In truth, engineering is

## SECTIONS

- •6.1 Society's View of Engineering
- •6.2 How Engineers Learn From Mistakes
- •6.3 The Role of Failure in Engineering Design: Case Studies
- •6.4 Preparing for Failure in Your Own Design

## OBJECTIVES

*In this chapter, you will*

- Examine society's view of the engineer
- Learn about the role of failure in engineering design
- Discuss classic design failures as case studies
- Learn how to accept and utilize failure along the path to engineering success

a glamorous profession that attracts countless creative individuals. Engineers have changed the world and improved the quality of life in more ways than can be counted. The reason the public views engineers with apprehension may be due to the following fact: *Engineers do the impossible as a matter of routine.* What engineers accomplish is so amazing that it can't be understood by the general public. Consider the debut of the original television series "Star Trek" in the late 1960s. The notion that Captain Kirk could take a small box out of his hip pocket, flip open the cover, and talk to anyone in the world was pure science fiction. Now this scenario is a reality. You can go out today, buy a tiny, relatively inexpensive cell phone that fits in your pocket, and call anyone in the world who has a telephone.

In the first half of the 1900s, much of science fiction focused on imaginary trips to the moon. Flash Gordon and Ming the Merciless captivated the minds of youngsters and adults. During the 1960s, engineers made it possible for the United States to *actually* land on the moon. Although the movie Apollo 13 highlighted the trying ordeal of brave astronauts in the face of an unexpected mishap, in truth it was the NASA engineers that safely brought the astronauts back to earth.

As an engineer-to-be, you will have a great responsibility to keep the public on track when it comes to popular misconceptions about engineering. When a public works project comes before your city government, it will be the engineers in town who will be able to discuss its impact on a sound technical basis. When the cell phone company wants to place a tower inside a church steeple, an engineer will be able to intelligently discuss the safety questions raised by the public. When the local school district needs help defining its strategic goal for information technology in the classroom, it will be the engineers as residents who will be most able to guide the planning with intelligence.

These insights do not come easily. The public harbors numerous misconceptions about the technical world that surrounds us. The following true events highlight the discord between the errant notions of the nontechnical public and the knowledge of the engineer.

*Cell phone use*: A newspaper in the Midwest ran a story about a father who got locked out of a tornado shelter during a storm. He had run out of the shelter to retrieve his cell phone from his truck so that his family would have communication if the phone lines went down in the town.

*Misconception*: Although the cell phone instruments themselves are wireless, the signals are sent to nearby cell-phone towers where they are routed thereafter to the telephone network by wires. In rural communities, these wires are most certainly overhead wires that would be vulnerable to a tornado.

*Magnetic resonance imaging*: The official name of magnetic resonance imaging (MRI) used to be "nuclear magnetic resonance" (NMR), having been named after the basic physical principle that lies at the core of this indispensable medical diagnostic tool. The name was changed to MRI because too many people were uncomfortable with anything involving nuclear technology.

*Misconception*: The "nuclear" in NMR refers to the nucleus of the atom being probed which is made to resonate in the presence of a magnetic field. The physics involved has nothing whatsoever to do with radioactivity of any kind.

*The debate over Napster*: The year 2001 saw a great public debate over the legality and morality of the free music download Web site *napster.com*. Amidst the discussion, an editorial appeared in a college newspaper noting that Napster aficionados obviously preferred the sound of MP3 digital music available over the Web to the sound of regular music available on CDs.

*Misconception*: CD and MP3 recordings are both stored in digital format. The only difference is the storage medium. When played over the same sound system, the two formats are indistinguishable.

*Radio advertisement for a Palm VII*: The Palm Pilot™ has become a favorite personal digital assistant for many individuals. The Palm VII model includes cell-phone like capability that forms a wireless link to a subscriber Web provider. An advertisement that aired in 2001 featured a wayward sole who gets locked in a meat freezer and uses his wireless Palm VII to send an e-mail message asking for help.

*Misconception*: Meat lockers are invariably made with solid metal walls and doors. It would be impossible for the wireless signals to enter or leave the meat locker.

*Making a good hot cup of tea*: A TV sitcom featured a grandmother who kept her teapot boiling on the stove because she wanted it "extra hot."

*Misconception*: Water boils at 100°C. Once it boils, it can become no hotter without turning to steam and escaping the pot.

*Crash of cell phone system*: The Avon 3-Day Breast Cancer walk is an annual spring fund-raising event that takes place in many cities around the United States. The last leg of the 60-mile walk is paced so that participants arrive together at the end point in a large closing ceremony. After the ceremony, participants must find their families or friends for transportation home. At the Boston 2000 event, which included about 2,500 walkers, the entire local cell network crashed after several hundred walkers and several hundred picker-uppers attempted to reach each other by cell phone.

*Misconception*: The traffic volume of any cell phone network is finite. Cell phones do not call directly from one to the other, but must be routed via a cell-phone tower to land-based links. The system has only a few dozen channel frequencies and local lines available from any one tower.

*Electric wiring*: A homeowner called an electrician complaining about a light that had a "short" and flickered from time to time.

*Misconception*: The lighting circuit most certainly had an intermittent *open* circuit, and not a *short*, or closed, circuit. As any engineering student would know, a short circuit formed by two opposite wires accidentally touching each other would provide an unobstructed path to electricity that would instantly trip a circuit break or fuse rather than allow the light to flicker.

*Thermostat Control*: A rapid response help line (911) advised a parent who had found a young child lost in the snow to turn up the cabin thermostat very high so that the room would heat up "very fast" while the family waited for a distant ambulance to arrive.

*Misconception*: Most heating systems have just two states: "on" and "off." The heater will turn on when the inside temperature falls below the thermostat temperature, but the *rate* at which the room heats up will depend solely on the furnace output and will be independent of the thermostat setting.

*Laser Printer*: Many people think that the common laser printer works by burning the letters in the page with a laser.

*Misconception*: The sole role of the laser in a laser printer is to condition a light-sensitive roller so that it will hold electrostatically-charged toner particles where printing is desired. These toner particles are then transferred to the paper and fused to it by a heated roller. The laser beam never makes contact with the paper. This process is the same on used in photocopiers ("xerox" machines) except that in the latter, a broad flash of bright light it used to condition the drum to hold toner particles in the desired places.

## 6.2    HOW ENGINEERS LEARN FROM MISTAKES

Consider the following scenario that describes the experience of two students working on the design of a battery-powered vehicle for a design competition:

> The students had been working on their vehicle for almost a week. Having decided to enter the design competition, they directed their efforts towards the design of an articulated arm that would pick up the opposing vehicle and throw it off the track. They first outlined their design on paper and then tested it out using computer-aided design (CAD) software available in the computer lab. They built all the parts in the school machine shop, using their CAD drawings as a guide. A sketch of their design is shown in Figure 6.1. The students had just finished putting together all 58 machined pieces and were delighted to find that the arm worked perfectly on the very first try!

Wouldn't it be nice if engineering were so simple and foolproof as the scenario depicted in the preceding paragraph? In the real world, almost nothing works correctly the first time. Getting things to work perfectly almost always takes longer than expected. Fabricated parts do not fit together, circuits have wiring errors, software modules have incompatibilities, and structural elements are incorrectly sized. Experienced engineers know that designs seldom work the first time and are never discouraged by initial failure.

Engineering tasks take longer than expected, because *failure* is an inevitable part of the design process. It's unreasonable to expect a new design to work the first time. When a device does not work as planned, it's a sure sign that something important has been overlooked. Perhaps two moving parts hit each other unexpectedly. Perhaps an electronic circuit or machine doesn't work because stray effects were not included in the design model. A software program may fail because an unforeseen set of keystrokes leads to a logical dead end. A scale model of a bridge may reveal an overstressed support beam because a support pillar was omitted. A biomedical implant may be rejected because a tissue interaction was underestimated. Whatever the failure mode, it's better for a defect to appear *during* the design process than after it, when the device is in the field.

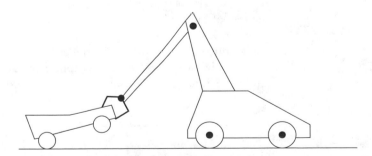

***Figure 6.1.***    Hypothetical, impossible-to-build car design includes an articulated arm.

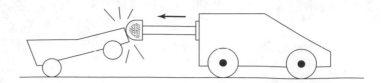

**Figure 6.2.** More realistic battering ram design as an offensive strategy.

Redesign after failure provides an important path to optimizing the final product and gives engineers needed time to correct deficiencies and work out bugs.[1]

A more realistic version of the paragraph appearing at the start of this section might resemble the following:

> The students had been working on their design competition vehicle for almost a month. Their first version of the car included an articulated arm designed to pick up the opposing vehicle and throw it off the track. After much trial and error, they succeeded in building an arm capable of lifting objects. At first, the students attempted to power the arm using the spring force from a single mousetrap, but after much testing, the students discovered that they had ignored frictional effects. Two mousetraps were required for adequate mechanical power. Eventually, they learned that powering the arm from a small electric motor and gear set provided much finer control of its movements. Their first attempt at assembly required that they return to the machine shop to redrill several holes they had put in the wrong places. Although the final version of the arm worked well on paper and also on a CAD program available in the computer lab, the arm failed miserably when mounted on the car. When an opposing program vehicle was lifted into the air, the center of gravity fell outside the maximum allowed wheelbase of the vehicle, causing *both* vehicles to topple over and fall off the track. The students had to abandon their arm design and eventually settled on the battering ram of Figure 6.2 for their offensive strategy.

## 6.3  THE ROLE OF FAILURE IN ENGINEERING DESIGN: CASE STUDIES

The pages of engineering history are full of examples of design flaws that escaped detection in the design phase only to reveal themselves once the device was in actual use. Although many devices are plagued by minor design flaws from time to time, a few failure cases have become notorious because they affected many people, caused great property damage, or led to sweeping changes in engineering practice. In this section, we review several design failures from the annals of engineering lore. Each event involved the loss of human life or major destruction of property, and each was caused by an engineering design failure. The mistakes were made by engineers who did the best they could, but had little prior experience or had major lapses in engineering judgement. After each incident, similar disasters were averted, because engineers were able to study the *causes* of the problems and establish new or revised engineering standards

---

[1] The term "bug" was coined by Adm. Grace Hopper in 1945. In the early days of computers, logic gates were made from electromechanical relay switches rather than transistors. A moth flew inside one such computer, got stuck between two relay contacts, and prevented seemingly closed relay contacts from making a true connection. The computer malfunctioned because of the "bug."

and guidelines. Studying these classic failures and the mistakes of the engineers who caused them will help you to avoid making similar mistakes in your own work.

The failure examples to follow all had dire consequences. Each occurred once the product was in use, long after the initial design, test, and evaluation phases. It's always better for problems to show up *before* the product has gone to market. Design problems can be corrected easily during testing, burn-in, and system evaluation. If a design flaw shows up in a product or system that has already been delivered for use, the consequences are far more serious. As you read the examples of this section, you might conclude that the causes of these failures in the field should have been obvious, and that failure to avoid them was the result of some engineer's carelessness. Indeed, it's relatively easy to play "Monday-morning quarterback" and analyze the cause of a failure *after* it has occurred. But as any experienced engineer will tell you, spotting a hidden flaw during the test phase is not always easy when a device or system is complex and has many parts or subsystems that interact in complicated ways. Even simple devices can be prone to hidden design flaws that elude the test and evaluation stages. Indeed, one of the marks of a good engineer is the ability to ferret out flaws and errors *before* the product finds its way to the end user. You can help to strengthen your abilities with the important intuitive skill of flaw detection by becoming familiar with the classic failure incidents discussed in this section. If you are interested in learning more details about any of the case studies, you might consult one of the references listed at the end of the chapter.

### 6.3.1   Case 1: Tacoma Narrows Bridge

The Tacoma Narrows Bridge, built across Puget Sound in Tacoma, Washington in 1940, was the longest suspension bridge of its day. The design engineers copied the structure of smaller, existing suspension bridges and simply built a longer one. As had been done with countless shorter spans, support trusses deep in the structure of the bridge's framework were omitted to make it more graceful and visually appealing. No calculations were done to prove the structural integrity of a longer bridge lacking internal support trusses. Because the tried-and-true design methods used on shorter spans had been well tested, the engineers assumed that these design methods would work on longer spans. On November 7, 1940, during a particularly windy day, the bridge started to undulate and twist, entering into the magnificent torsional motion shown in Figure 6.3. After several hours, the bridge crumbled as if it were made from dry clay; not a piece remained between the two main center spans.

What went wrong? The engineers responsible for building the bridge had relied on calculations made for smaller bridges, even though the assumptions behind those calculations did not apply to the longer span of the Tacoma Narrows Bridge. Had the engineers heeded some basic scientific intuition, they would have realized that three-dimensional structures cannot be directly scaled upward without limits.

### 6.3.2   Case 2: Hartford Civic Center

The Hartford Civic Center was the first of its kind. At the time of its construction in the mid 1970s, no similar building had been built before. Its roof was made from a space frame structure of interconnected rods and ball sockets, much like a child's construction toy. Hundreds of rods were interconnected in a visually appealing geodesic pattern like the one shown in Figure 6.4. Instead of performing detailed hand calculations, the design engineers relied on the latest computer models to compute the loading on each individual member of the roof structure. Recall that computers in those days were much more primitive than those we enjoy today. The PC had not yet been invented, and all work was performed on slow large-mainframe computers.

***Figure 6.3.*** The Tacoma Narrows Bridge in torsional vibration.

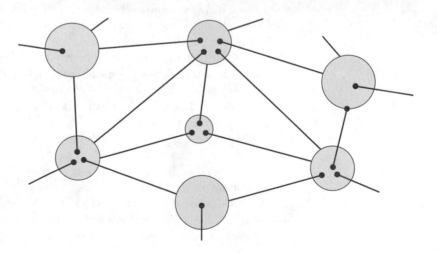

***Figure 6.4.*** Geodesic, rod-and-ball socket construction.

On January 18, 1978, just a few hours after the center had been filled to capacity with thousands of people watching a basketball game, the roof collapsed under a heavy snow load, demolishing the building. Miraculously, no one was hurt in the collapse.

Why did the collapse occur? Some attribute the failure to the engineers who designed the civic center and chose not to rely on their basic judgement and intuition gleaned from years of construction practice. Instead, they relied on computer models of their new space frame design. These computer models had been written by programmers, not structural engineers, during the days when computer modeling was in its infancy. The programmers based their code algorithms on structural formulas from textbooks. Not one of the programmers had ever actually built a roof truss. All failed to include basic derating factors at the structural joints to account for the slight changes in layout (e.g., minor variations in angles, lengths, and torsion) that occur when a complex structure is actually built. The design engineers trusted the output of computer models that never had been fully tested on actual construction. Under normal roof load, many ball-and-socket joints were stressed beyond their calculated limits. The addition of a heavy snow load to the roof load proved too much for the structure to bear.

### 6.3.3   Case 3: Space Shuttle *Challenger*

The NASA Space Shuttle *Challenger* blew up during launch on a cold day in January 1986 at Cape Kennedy (Canaveral) in Florida. Thousands witnessed the explosion as it happened [see Figure 6.5]. Hundreds of millions watched news tapes of the event for weeks afterwards. After months of investigation, NASA traced the problem to a set of O-rings used to seal sections of the multisegmented booster rockets. The seals were never designed to be operated in cold weather, and on that particular day, it was about 28°F (−2°C), a very cold day for Florida. The frozen O-rings were either too stiff toproperly seal the sections of the booster rocket or became brittle and cracked due to the unusually cold temperatures. Flames spewed from an open seal during acceleration and ignited an adjacent fuel tank. The entire spacecraft blew up, killing all seven astronauts on board, including a high school teacher. It was the worst space disaster in U.S. history.

**Figure 6.5.**   The Space Shuttle *Challenger* explodes during launch. (*Photo courtesy of RJS Associates.*)

In using O-rings to seal adjacent cylindrical surfaces, such as those depicted in Figure 6.6, the engineers had relied on a standard design technique for rockets. The Challenger's booster rockets, however, were much larger than any on which O-rings had been used before. This factor, combined with the unusually cold temperature, brought the seal to its limit, and it failed.

There was, however, another dimension to the failure. *Why* had the booster been built in multiple sections, requiring O-rings in the first place? The answer is complex, but the cause was largely attributable to one factor: The decision to build a multisection booster was, in part, *political.* Had engineering common sense been the sole factor, the boosters would have been built in one piece without O-rings. Joints are notoriously weak spots, and a solid body is almost always stronger than a comparable one assembled from sections. The manufacturing technology existed to build large, one-piece rockets of appropriate size. But a senator from Utah lobbied heavily to have the contract for constructing the booster rockets awarded to a company in his state. It was not physically possible to transport a large, one-piece booster rocket all the way from Utah to Florida over existing rail lines. Trucks were too small, and no ships were available that could sail to land-locked Utah, which lies in the middle of the United States. The decision by NASA to award the contract to the Utah company resulted in a multisection, O-ring-sealed booster rocket whose smaller pieces would easily be shipped by rail or truck.

Some say the catastrophe resulted from a lack of ethics on the part of the design engineers who suspected the O-ring design of having potential problems. Some say it was the fault of NASA for succumbing to political pressure from Congress, its ultimate funding source. Others say it was just an unusual convergence of circumstances, since neither the Utah senator nor the design engineers knowingly advocated for a substandard product. The sectioned booster had worked flawlessly on many previous shuttle flights that had not been launched in subfreezing temperatures. Still, others say that by putting more weight on a political element of the project, rather than on pure engineering concerns, the engineers were compromised into a less-than-desirable design concept that had never before been attempted on something so large.

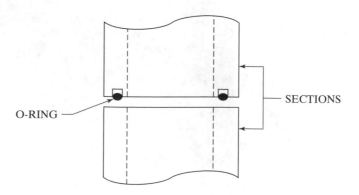

**Figure 6.6.** Schematic depiction of O-ring seals.

### 6.3.4   Case 4: Kansas City Hyatt

If you've ever been inside a Hyatt hotel, you know that their internal architectures are very unique. The typical Hyatt hotel has cantilevered floors that form an inner trapezoidal atrium, and the walkways and halls are open, inviting structures. There's nothing quite like the inside of a Hyatt. In the case of the Kansas City Hyatt, first opened in 1981, the design included a two-layer, open-air walkway that spanned the entire lobby in midair, from one balcony to another. During a party that took place not long after the hotel opened, the walkway was filled with people dancing in time to the music. The weight and rhythm of the load of people, perhaps in resonance with the walkway, caused it to collapse suddenly. Over one hundred people died, and the event will be remembered forever in the history of hotel management. Although the hotel eventually reopened, to this day the walkway has never been rebuilt.

The collapse of the Hyatt walkway is a classic example of failure due to lack of construction experience. In this case, however, the error originated during the *design* phase, not the construction phase. In order to explain how the walkway collapsed, consider the sketch of the skeletal frame of the walkway, as specified by the design engineer, shown here in Figure 6.7.

Each box beam was to be held up by a separate nut threaded onto a suspended steel rod. The rated load for each nut-to-beam joint was intended to be above the maximum weight encountered during the time of the accident. What's wrong with this picture? The problem is that the structure as specified was not a realistic structure to build. The design called for the walkway's two decks to be hung from the ceiling by a single rod at each support point. The rods were made from smooth steel having no threads. Threading reduces the diameter of a rod, so it's impossible to get a nut to the middle of a rod unless the rod is threaded for at least half its length. In order to construct the walkway as specified, each rod would have to be threaded along about 20 feet of its length, and numerous rods were needed for the long span of the walkway. Even with an electric threading machine, it would have taken days to thread all the needed rods. The contractor who actually built the walkway proposed a modification to the construction so that only the very ends of the rods would have to be threaded. The modification is illustrated in Figure 6.8.

The problem with this modification is that the nut (A) at the lower end of the upper rod now had to support the weight of *both* walkways. A good analogy would be two mountain climbers hanging onto a rope. If both grabbed the rope simultaneously, but independently, the rope could hold their weight. If the lower climber grabbed the ankles of the upper climber instead of the rope, however, the upper climber's hands would have to hold the weight of *two* climbers. Under the full, or maybe excessive, load conditions of that day, the weight on nut (A) of the Hyatt walkway was just too much, and the joint gave way. Once the joint on one rod failed, the complete collapse of the rest of the joints and the entire walkway quickly followed.

Some attributed the fatal flaw to the senior design engineer who specified single rods requiring 20 feet of threading. Others blamed it on the junior engineer, who signed off on the modifications presented by the construction crew at the construction site, and the senior engineer, who should have communicated to the junior engineer the critical nature of the rod structure as specified. Perhaps both engineers lacked seasoning—the process of getting their hands dirty on real construction problems as a way of gaining a feeling for how things are made in the real world.

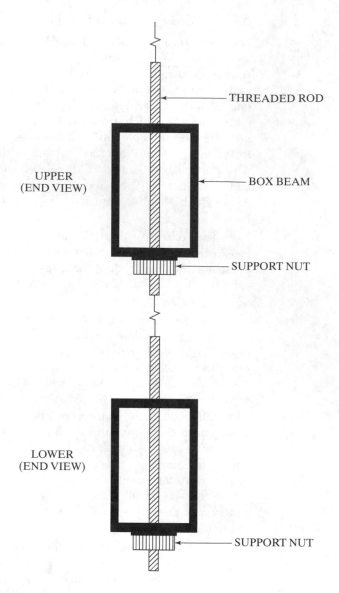

THREADED ROD

UPPER
(END VIEW)

BOX BEAM

SUPPORT NUT

LOWER
(END VIEW)

SUPPORT NUT

**Figure 6.7.**    Kansas City Hyatt walkway support structure as designed.

Regardless of who was at fault, the design also left little room for *safety margins*. It's common practice in structural design to leave *at least* a factor-of-two safety margin between the calculated maximum load and the expected maximum load on a structure. The safety margin allows for inaccuracies in load calculations due to approximation, random variations in material strengths, and small errors in fabrication. Had the walkway included a safety margin of a factor of two or more, the doubly stressed joint on the walkway might not have collapsed, even given its modified construction. The design engineers specified a walkway structure that was possible, but not practical, to build. The construction supervisor, unaware of the structural implications, but wishing to see the job to completion, ordered a small, seemingly innocent, but ultimately fatal, change in the construction method. Had but one of the design engineers ever spent time working on a construction site, this shortcoming might have been discovered. Errors such as

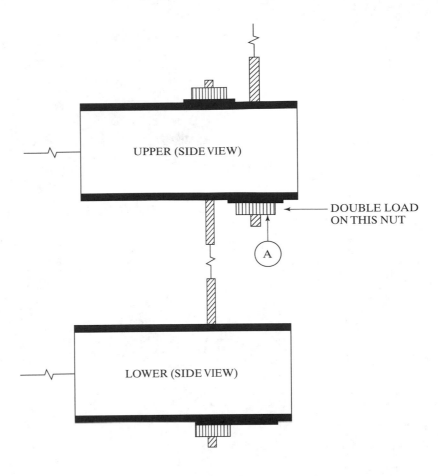

**Figure 6.8.** Kansas City Hyatt walkway support structure as actually built.

the one that occurred at the Kansas City Hyatt can be prevented by including workers from all phases of construction in the design process, ensuring adequate communication between all levels of employees, and adding far more than minimal safety margins where public safety is at risk.

## 6.3.5  Case 5: Three Mile Island

Three Mile Island was a large nuclear power plant in Pennsylvania (see Figure 6.9). It was the sight of the worst nuclear accident in the United States and nearly comparable to the total meltdown at Chernobyl, Ukraine. Fortunately, the incident at Three Mile Island resulted in only a near miss at a meltdown, but it also led to the shutting down and trashing of a billion-dollar electric power plant and significant loss of electrical generation capacity on the power grid in the eastern United States.

On the day of the accident, a pressure buildup occurred inside the reactor vessel. It was normal procedure to open a relief valve in such situations to reduce the pressure to safe levels. The valve in question was held closed by a spring and was opened by applying voltage to an electromagnetic actuator. The designer of the electrical control system had made one critical mistake. As suggested by the schematic diagram shown in Figure 6.10, indicator lights in the control room lit up when power was applied to or removed from the valve actuator coil, but the control panel gave no indication about the

**Figure 6.9.** Three Mile Island power plant.

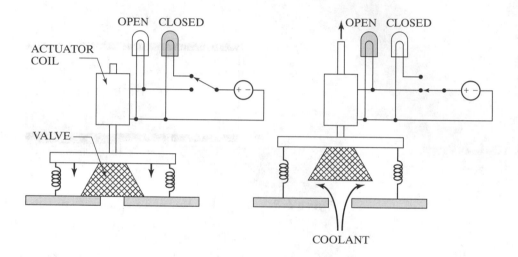

**Figure 6.10.** Valve indicator system as actually designed.

the *actual* position of the valve. After a pressure-relief operation, the valve at Three Mile Island became stuck in the open position. Although the actuation voltage had been turned off and lights in the control room indicated the valve to be closed, it was actually stuck open. The mechanical spring responsible for closing the valve did not have enough force to overcome the sticking force. While the operators, believing the valve to be closed, tried to diagnose the problem, coolant leaked from the vessel for almost two hours. Had the operators known that the valve was open, they could have closed it manually or taken other corrective measures. In the panic that followed, however, the operators continually believed their control-panel indicator lights and thought that the valve was closed. Eventually the problem was contained, but not before a rupture nearly occurred in the vessel. Such an event would have resulted in a complete core meltdown and spewed radioactive gas into the atmosphere. Even so, damage to the reactor core was so severe that the plant was permanently shut down. It has never reopened.

The valve actuation system at Three Mile Island was designed with a poor human–machine interface. The ultimate test of such a system, of course, would be

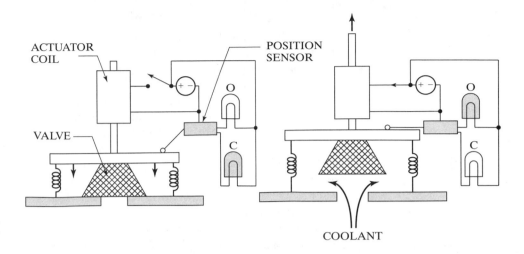

**Figure 6.11.**    Valve indicator system as it should have been designed and built.

during an emergency when the need for absolutely accurate information would be critical. The operators assumed that the information they were receiving was accurate, while in reality it was not. The power plant's control panel provided the key information by inference, rather than by direct confirmation. A better design would have been one that included an independent sensor that unambiguously verified the true position of the valve, as suggested by the diagram of Figure 6.11.

### 6.3.6    Case 6: USS Vincennes

The Vincennes was a U.S. missile cruiser stationed in the Persian Gulf during the Iran-Iraq war. On July 3, 1988, while patrolling the Persian Gulf, the Vincennes received two IFF (Identification: Friend or Foe) signals on its Aegis air-defense system. Aegis was the Navy's complex, billion-dollar, state-of-the-art information-processing system that displayed more information than any one operator could possibly hope to digest. Information saturation was commonplace among operators of the Aegis system. The Vincennes had received two IFF signals, one for a civilian plane and the other for a military plane. Under the pressure of anticipating a possible attack, the overstimulated operator misread the cluttered radar display and concluded that only one airplane was approaching the Vincennes. Repeated attempts to reach the nonexistent warplane by radio failed. The captain concluded that his ship was under attack and made the split-second decision to have the civilian airplane shot down. Two hundred ninety civilians died needlessly.

What caused this catastrophic outcome? Was it bad military judgment? Was it an operating error? Were the engineers who designed the system at fault? The Navy officially attributed the accident to "operator error" by an enlisted sailor, but in some circles the blame was placed on the engineers who had designed the system. Under the stress of possible attack and deluged with information, the operator simply could not cope with an ill-conceived human–machine interface designed by engineers. Critical information, being needed most during crisis situations, should have been uncluttered and easy to interpret. The complex display of the Aegis system was an example of something that was designed just because it was technically possible. It resulted in a human–machine interface that became a weak link in the system.

### 6.3.7    Case 7: Hubble Telescope

The Hubble is an orbiting telescope that was put into space at a cost of over a billion dollars. Unaffected by the distortion experienced by ground-based telescopes due to atmospheric turbulence, the Hubble has provided spectacular photos of space and has made possible numerous astronomical discoveries. Yet the Hubble telescope did not escape design flaws. Of the many problems that plagued the Hubble during its first few years, the most famous was its improperly fabricated mirrors. They were distorted and had to be corrected by the installation of an adaptive optic mirror that compensated for aberrations. The repairs were carried out by a NASA Space Shuttle crew. Although this particular flaw is the one most often associated with the Hubble, it was attributed to sloppy mirror fabrication rather than to a design error. Another, less-well-known design error more closely illustrates the lessons of this chapter. The Hubble's solar panels were deployed in the environment of space, where they were subjected to alternate heating and cooling as the telescope moved in and out of the earth's shadow. The resulting expansion and contraction cycles caused the solar panels to flap like the wings of a bird. Attempts to compensate for the unexpected motion by the spacecraft's computer-controlled stabilizing program led to a positive feedback effect which only made the problem worse. Had the design engineers anticipated the environment in which the telescope was to be operated, they could have compensated for the heating and cooling cycles and avoided the problem. This example illustrates that it's difficult to anticipate all the conditions under which a device or system may be operated. Nevertheless, extremes in operating environment often are responsible for engineering failures. Engineers must compensate for this problem by testing and *retesting* devices under different temperatures, load conditions, operating environments, and weather conditions. Whenever possible (though obviously not possible in the case of the Hubble), a system should be developed and tested in as many different environmental conditions as possible if a chance exists that those conditions will be encountered in the field.

### 6.3.8    Case 8: De Haviland Comet

The De Haviland Comet was the first commercial passenger jet aircraft. A British design, the Comet enjoyed many months of trouble-free flying in the 1950s until several went down in unexplained crashes. Investigations of the wreckages suggested that the fuselages of these planes had ripped apart in midflight. For years, the engineers assigned the task of determining the cause of the crashes were baffled. What, short of an explosion, could have caused the fuselage of an aircraft to blow apart in flight? No evidence of sabotage was found at any of the wreckage sites. After some time, the cause of the crashes was discovered. No one had foreseen the effects of the numerous pressurization and depressurization cycles that were an inevitable consequence of takeoffs and landings. Before jet aircraft, lower altitude airplanes were not routinely operated under pressure. Higher altitude jet travel brought with it the need to pressurize the cabin. In the case of the Comet, the locations of the rivets holding in the windows developed fatigue cracks, which, after many pressurization and depressurization cycles, grew into large, full-blown cracks in the fuselage. This mode of failure is depicted in Figure 6.12.

Had the design engineers thought about the environment under which the finished product would be used, the problem could have been avoided. Content instead with laboratory stress tests that did not mimic the actual pressurization and depressurization cycles, the engineers were lulled into a false sense of security about the soundness of their design. This example of failure again underscores an important engineering lesson: Always test a design under the most realistic conditions possible. Always assume that environmental conditions will affect performance and reliability.

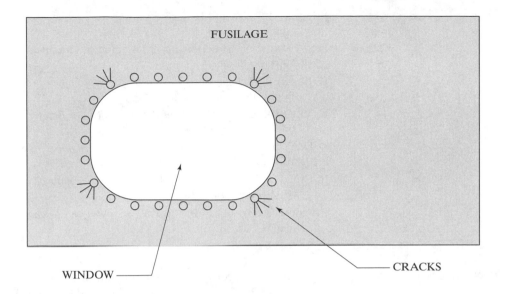

**Figure 6.12.**    Stress cracks around the window rivets of the De Haviland Comet.

## 6.4    PREPARING FOR FAILURE IN YOUR OWN DESIGN

First-time designs often betray previously hidden flaws after an initial period of successful use. Design flaws eventually show up because the operating environment changes, a previously untried sequence of events occurs, or weak points in the design encounter repeated stress. Sometimes, failure occurs just because of plain old statistics. As the saying goes, "If something is bound to fail, it *will* fail sooner or later." After a failure occurs, it's the engineer's job to determine the cause, fix what's wrong, and begin tests again. At the same time, it's up to the design engineer to identify as many of the bugs and weaknesses as possible during the early phases of the design process. Thorough testing and retesting under all sorts of operating conditions is essential. An unsuccessful first prototype presents an excellent opportunity to discover and weed out bugs before the final version of the device is put to market.

In the commercial sector, the rush to bring a product to market ahead of competitors puts pressure on the engineer to complete the test and evaluation phases as quickly as possible. For this reason, many consumer products, including automobiles, computers, and software, develop problems soon after they are released. If you purchase one of these items during the first year of issue, expect to find bugs and weaknesses that were not discovered on the factory floor.

Despite this admonition, you should be eager to apply your engineering design skills to new technology and innovation. If all engineers were content to stay with tried-and-true designs, technical progress would cease. Understanding when to stay within the bounds of a traditional design and when to move on to new creative frontiers requires experience, knowledge, and intuition. When you encounter failure in your own design projects, do not be discouraged. Recognize failure as an inevitable part of the design process. Use it to learn, discover, and expand your capabilities as an engineer.

## KEY TERMS

Failure                          Flaw                          Bugs
Safety margin

**REFERENCES**

J. FELD, KENNETH and L. Carper, *Construction Failure—Wiley Series of Practical Construction Guides,* New York: John Wiley & Sons, 1996.

E. S. FERGUSON, "How Engineers Lose Touch," *Invention and Technology.* Vol. 8 (3), Winter 1993, pp.16–24.

H. PETROSKI, *To Engineer is Human: The Role of Failure in Successful Design.* New York: Vintage Books, 1992.

H. PETROSKI, *Design Paradigms: Case Histories of Error and Judgement in Engineering.* Cambridge: Cambridge Univ. Press, 1994.

D. D. A. PIESOLD, *Civil Engineering Practice: Engineering Success by Analysis of Failure.* New York: McGraw-Hill, 1991.

R. UHL, G. M. DAVIDSON, and K. A. ESAKLUL, *Handbook of Case Histories in Failure Analysis.* ASM International, 1992.

P. VISWANADHAM and P. SINGH, *Failure Modes and Mechanisms in Electronic Packages.* Chapman & Hall, 1997.

**PROBLEMS**

1. Identify a product or system in your own experience that has failed. Write a short summary of the cause of the failure and how you might improve upon the design.

Look up and write a synopsis of the following classic engineering failure incidents:

2. Exxon Bayway refinery, Linden, New Jersey (1990)
3. General Electric rotary compressor refrigerators (1990)
4. Green Bank radio telescope (1989)
5. Union Carbide chemical leak, Bhopal, India (1984)
6. Korean Airlines Flight 007 (1983)
7. Interstate 95 Bridge, Mianus River, Connecticut (1983)
8. Alexander L. Kielland oil platform, North Sea (1980)
9. American Airlines DC-10 (1979)
10. Skylab (1979)
11. The New York City Power Blackout (1976)
12. The windows in the John Hancock Tower, Boston, Massachusetts (1976)
13. Big Ben, London (1976)
14. Bay Area Rapid Transit (BART) (1973)
15. Point Pleasant Bridge, Ohio River, Ohio-West Virginia (1967)
16. The Apollo 1 capsule fire (1967)
17. The Great Northeast Power Blackout (1965)
18. Quebec City Bridge (1907)
19. The Johnstown flood (1889)
20. Liberty Bell, Philadelphia (1835)

# 7

# Learning to Speak, Write, and Make Presentations

Imagine that you have just purchased a new, top-of-the-line computer with the fastest available processor, many megabytes of memory, a huge disk drive, and a large, sophisticated sound card. When you get the computer home from the store, you notice that the model you bought has been shipped without a monitor. In fact, the computer is missing entirely a socket where a monitor can be connected. In all other respects, the computer is state-of-the-art and in perfect condition. What would your reaction be? You probably would take the computer back to the store claiming, of course, that any computer, no matter how fast or powerful, is useless without a means to extract the information it produces. Now imagine a similar scenario in which the computer has a monitor but comes with a very poor modem capable of a maximum data transfer rate of only 300 baud (about 30 printed characters per second). Such a slow modem would indeed limit the efficiency with which your computer could talk to other computers over the Internet. Reading electronic mail (e-mail) would be hopelessly slow, and accessing Web pages would be next to impossible. You probably would take this second computer back to the store also, claiming it to be a powerful machine that is incapable of communicating with others.

## SECTIONS

- 7.1 The Importance of Good Communication Skills
- 7.2 Preparing for Meetings, Presentations, and Conferences
- 7.3 Preparing for a Formal Presentation
- 7.4 Writing Electronic Mail, Letters, and Memoranda
- 7.5 Writing Technical Reports, Proposals, and Journal Articles
- 7.6 Preparing an Instruction Manual
- 7.7 Producing Good Technical Documents: A Strategy

## OBJECTIVES

*In this chapter, you will learn about*

- Writing effective e-mail messages and memos.
- Preparing for meetings, presentations, and conferences.
- Writing technical proposals, reports, and journal papers.
- Preparing an instruction manual.
- Identifying the characteristics of good oral and written communication.
- Studying examples of good and bad writing styles.

People are a little bit like computers in this respect. The smartest person in the world, the fastest thinker, the most prolific scientist or engineer, is severely handicapped without an ability to communicate with others. The famous physicist Dr. Steven Hawking, author of *A Brief History of Time*[1] and other works on cosmology, has been acknowledged as one of the most brilliant minds of modern physics. A debilitating disease known as ALS took away his ability to speak and move his limbs and facial muscles, cutting off all normal means of communication. His thoughts and ideas have come to the world one agonizing word at a time by way of synthesized computer speech. How much more the world would understand about the nature of the universe were Stephen Hawking able to communicate using normal human speech and body language

## 7.1 THE IMPORTANCE OF GOOD COMMUNICATION SKILLS

Asked what the most important single skill is for new engineers in the workplace, most every employer will state emphatically: *communication skills!* (Figure 7.1). The best software programmer in the world will do a poor job if the ideas behind the software are not well understood by the user. The most proficient mechanical designer will fail if the structure being designed is set up incorrectly or is used in an improper manner. Communication skills are so important that the Accreditation Board for Engineering and Technology, the national organization responsible for accrediting programs in schools of engineering in the U.S. and Canada, has listed oral and written communication as mandatory elements for all engineering programs regardless of discipline. Listening also is a very important skill for students of all types. Engineering departments have risen to the challenge by teaching these skills in many different ways, from writing workshops and courses taught by English departments to required oral presentations by students in key core courses.

In this chapter, we cover some of the most basic of oral and written communication skills. Although not all inclusive, the scenarios presented in this chapter cover many of the communication situations in which you might find yourself during your engineering career. Topics include preparing for meetings, conferences and presentations, drafting short memos, writing letters and electronic mail, and writing long technical reports and journal papers.

## 7.2 PREPARING FOR MEETINGS, PRESENTATIONS, AND CONFERENCES

Whenever you get together to socialize, you are participating in an informal meeting. You also participate in an informal meeting when you meet with your boss or co-workers to discuss the status of your latest project. Are these two events similar? Should you approach both with the same level of preparation? In the first case, you'd feel silly if you prepared beforehand for a social gathering with your friends. The purpose of such a gathering would be to relax and dispense with formalities. In the case of a meeting with your boss, you *should* spend time preparing beforehand because the meeting will reflect upon your work, your competency, and your role and status within the company. The same could be said for a meeting with your professor to discuss your work as a student. Preparing for an informal meeting takes only a small amount of forethought and planning, because no long speech or formal presentation is required. On the contrary, you may come across as being phony if your conversation seems contrived or orchestrated.

---

[1] Stephen W. Hawking, *A Brief History of Time.* Toronto: Bantam Books, 1988.

Natural speech and hand gestures are always preferable at an informal meeting. You should, however, take the time to think about the content of the meeting beforehand. What will be the topics of conversation? What are your own opinions or thoughts about those topics? Are you being called upon to provide information? If so, take the time to prepare a list that highlights your important points or gives a summary of recent data. Perhaps a one-page outline of your progress to date would be beneficial. Or maybe a list of future planned tasks would be appropriate. Whatever the setting, a brief, to-the-point, one-page document is always helpful at an informal meeting that isn't a social gathering. The following list gives a few examples of informal meetings of various types along with suggestions for a document that might be passed out to all those present at the meeting.

- Project status review: Prepare a one-page bullet list of accomplishments that you've achieved since the last meeting.

- Report on recent tests: Prepare a one- or two-page table showing test results.

- Discussion on market potential of customers: Compile a list of the 10 most important customers from the past 5 years.

- Product design review: Write a one-page summary of the key features of your design concept for the product under development.

- Changes in company procedure: Draw an outline of your proposed organizational chart.

**Figure 7.1.**   Good communication skills are important to all aspects of engineering.

## 7.3    PREPARING FOR A FORMAL PRESENTATION

Engineers are called upon frequently to make formal presentations to design teams, management personnel, customers, and other large groups of people. Sometimes, engineers must present formal papers at technical conferences. In contrast with an informal presentation, a formal presentation requires careful preparation. You should organize your talk much like a written document. It should contain an introduction, body, summary, and recommendations or conclusion. The following points will help you to prepare an interesting talk that meets its objectives:

1.  Know your audience and plan upon an appropriate level of detail.

2.  Assume that the customer is hearing about your topic for the first time.

3.  Check out audio-visual equipment before the audience arrives. If you are using a laptop computer to present a slide show, determine *ahead of time* that it interfaces properly with the room's computer projection equipment.

4.  Dress in suitable attire.

5.  Cite the purpose of the talk within the first few minutes.

6.  Tell your audience why it's you who is speaking.

7.  Show an outline of your talk at the beginning. Give an initial overview of what the presentation will address. A single anecdote can help put the audience at ease and establish rapport.

8.  Keep your talk simple. It's easy to get lost in technical details without addressing the main points of your presentation. Leave discussion of details for audience questions. In that way, you'll be able to spotlight only those technical issues of true interest to the audience.

9.  Keep your talk short. If you plan to use between 50 percent and 60 percent of the allotted time, you probably will end on time.

10. Ask your own questions, and then answer them. Prepare visual aids to use as your own cues. Use bullets (•) as thought initiators and breakpoints. Use your visual aides, calling attention to them as your talk progresses. Try to format all slides the same way. Maintain eye contact. Never read! If you've brought along notes, refer to them as infrequently as possible.

11. Do not show the audience equations. The audience has little time to decipher them. This rule can be broken occasionally.

12. End your talk with "thank you, any questions?" or a concluding slide. In this way, the audience will know when your talk is over.

13. Restate postpresentation questions so that the entire audience can hear them. Restating a question also will help to clarify its content and will give you an extra moment to formulate your response.

The following example illustrates the elements of a good oral presentation. The context is an engineer who is presenting the results of recent mechanical loading tests to a group of professors and students.

**EXAMPLE 7.1: MECHANICAL LOADING AND TESTING**

Dan began with a short introduction that explained the purpose of his talk. "Thank you all for coming to my presentation. In this talk I will summarize test results on samples of materials that are candidates for the chassis frame of our new vehicle. As you all know, we've decided to go with composites—matrices of glass, carbon fiber, and epoxy resin—for the principal structural members of the vehicle. This choice will result in some extra costs, because these materials are more expensive than aluminum, plastic, or steel, but ultimately we'll have a better performing and more competitive vehicle. I've done some initial tests on sample compositions and wish to share them with you." Dan displayed the first of his overhead transparencies:

Loading Tests on L-type Carbon Composites

Daniel Little

Electric Car Project

Team Xebec

His next slide summarized the content of his presentation, providing an overview to the audience:

**Summary of Presentation**

- Description of Composite Materials
- Selection of Test Samples
- Stress–Strain Properties (Nondestructive Test)
- Maximum Load to Failure (Destructive Test)
- Fatigue Tests (Destructive Test)

"Let me begin by providing a brief review of composites. These materials were invented by the aircraft industry as possible lightweight alternatives to more expensive metals, such as titanium and magnesium. Now they're used in everything from bicycles to sailboat masts. Composites are made from a matrix of carbon or glass threads woven into the desired shape and impregnated with high tensile-strength epoxy resin." Dan passed out several samples of composite materials. They were black in color and felt like plastic. He put up a slide that asked two questions he planned to answer during his talk:

**Structural Frame for the Peak-Performance Design Competition**

- Should we use composites?
- What composition of fiber and epoxy is best?

"Composition, to remind you, is a measure of the percent weight of carbon fiber to epoxy. The diameter of the fibers is also a factor. Basically, the more fiber there is in the mixture, the more expensive the material will be. The trick is to find the optimal composition while considering strength *and* cost. I have here some data on carbon composition, because carbon fiber is the type we're most likely to choose." Dan displayed an overhead that described the various composites he had tested:

"Above about a 24-percent fill rate, this particular composite has too little epoxy resin to hold together well," explained Dan, "and below 8 percent, too little carbon fiber to retain tensile strength."

Dan next described the first of the tests performed on the samples. "The first test consisted of determining the stress–strain relationship for each of the materials in the sample list for a number of different fiber diameters. As a review, let me remind you that *stress* is a fancy word for applied force, and *strain* is another term for the amount by which the material stretches (or compresses) in response to the applied force. In a *linear* material, the strain, or stretch, is directly proportional to the applied stress. Double the stress, and you've doubled the strain. If too much strain occurs due to too much applied stress, the material will go into its nonlinear region in which added stress produces almost no additional strain.

"Now let me explain the tests I've performed. The first involves measuring the stress–strain, or force–displacement, properties of the material using the following setup." Dan put up the overhead shown in Figure 7.2.

"The sample to be tested is first machined into a round bar of 4-mm cross section. It's then installed in a tensile test machine that can apply a stretching force to the bar. The applied force is measured by a load cell that produces a voltage in proportion to the applied force. A strain gage is used to measure the strain. It's actually a thin-film resistor that's glued right onto the side of the test sample. Its resistance changes in proportion to how much the sample has been stretched from its rest length. Here's a typical plot of data taken on one sample, in this case L-8." Daniel put up the slide shown in Figure 7.3.

"The slope of the linear portion of this curve will be equal to the reciprocal of the material's elastic constant. The breakpoint in the curve defines the elastic limit. For our application, we'd like an elastic constant of about 6 kN/mm and an elastic limit of at least 400 N to ensure an adequate safety margin of about five times the largest force we realistically expect on the components.

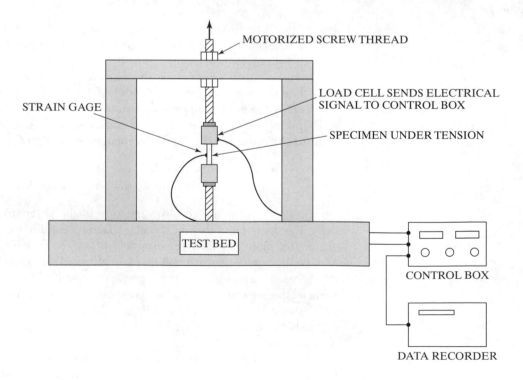

**Figure 7.2.** Instron™ test bed and computerized data recorder. The tensile test specimen is inside.

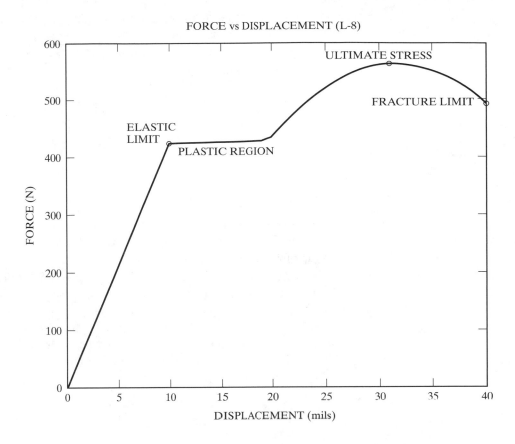

**Figure 7.3.** Force–displacement curve showing linear region and elastic limit.

**TABLE 7-1** Description of Various Composite Materials

| PRODUCT | PERCENT CARBON | APPROXIMATE COST PER POUND |
|---------|----------------|----------------------------|
| L-8 | 8 | $ 96 |
| L-10 | 10 | $ 102 |
| L-12 | 12 | $ 113 |
| L-16 | 16 | $ 120 |
| L-20 | 20 | $ 136 |
| L-24 | 24 | $ 141 |

"Let me show you the results of measurements obtained from one of my test matrices."[2] Dan put Table 7-2 up on the screen. He'd copied it from a page in his engineer's logbook, which served as a written record of the data.

"The entries in this first test matrix table indicate the measured elastic constant of each sample in kilonewtons per millimeter of displacement. This next table gives the elastic limit of each sample in kilonewtons." Dan then presented a slide showing Table 7-3.

---

[2] A *test matrix* consists of a table of methodically performed tests in which one variable changes on the vertical axis and one along the horizontal axis.

**TABLE 7-2** Measured Elastic Constant (kN/mm) Under Tensile Force for 4-mm Diameter Carbon Composite Test Samples

| | CARBON FIBER DIAMETER (MILS) | | | | |
|---|---|---|---|---|---|
| **SAMPLE** | **3** | **4** | **5** | **6** | **PERCENT CARBON** |
| L-8 | 8.4 | 7.1 | 6.5 | 6.0 | 8 |
| L-10 | 9.0 | 7.6 | 6.6 | 6.2 | 10 |
| L-12 | 9.5 | 8.2 | 6.9 | 6.5 | 12 |
| L-16 | 9.7 | 8.8 | 7.2 | 6.8 | 16 |
| L-20 | 9.9 | 9.2 | 7.5 | 7.1 | 20 |
| L-24 | 10.1 | 9.7 | 7.8 | 7.4 | 24 |

In the following figure, I've plotted the most important parameter, the elastic constant, versus percent carbon using fiber diameter as the third parametric variable." Daniel put up the plot shown in Figure 7.4.

"In my next overhead, I've plotted the elastic limit versus percent carbon, again using fiber diameter as the parametric variable." Dan put up the second plot of Figure 7.5.

"Because we need an elastic limit of at least 0.4 kN, we're limited to fiber diameters of 5 mils or more. Similarly, the largest elastic constants occur for 6-mil fiber, but the 5-mil fiber has adequate properties. Based on my tests, I'm recommending that we go with 16 percent carbon composite with 5-mil fiber for the prototype version of our vehicle. Thank you for your attention. Are there any questions?"

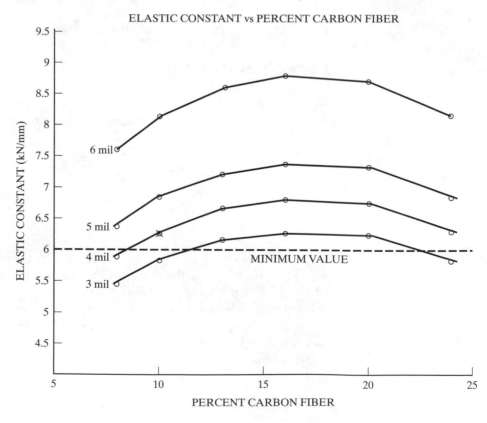

**Figure 7.4.** Force–displacement ratio versus percent carbon.

**TABLE 7-3**   Measured Elastic Limit (kN) Under Tensile Force for 4-mm Diameter Carbon Composite Test Samples

| | CARBON FIBER DIAMETER (MILS) | | | | |
| SAMPLE | 3 | 4 | 5 | 6 | PERCENT CARBON |
| --- | --- | --- | --- | --- | --- |
| L-8 | 0.28 | 0.33 | 0.41 | 0.52 | 8 |
| L-10 | 0.29 | 0.34 | 0.42 | 0.53 | 10 |
| L-12 | 0.30 | 0.36 | 0.43 | 0.56 | 12 |
| L-16 | 0.32 | 0.38 | 0.46 | 0.58 | 16 |
| L-20 | 0.30 | 0.36 | 0.44 | 0.56 | 20 |
| L-24 | 0.29 | 0.34 | 0.42 | 0.53 | 24 |

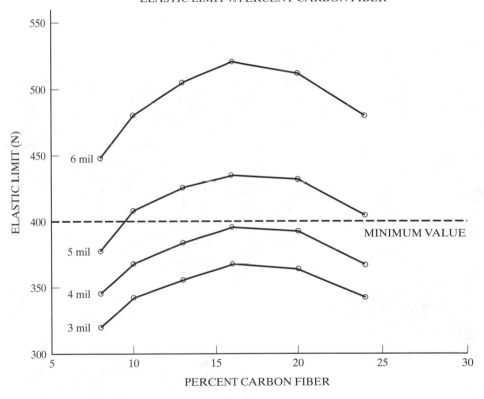

**Figure 7.5.**   Elastic limit versus percent carbon.

## 7.4   WRITING ELECTRONIC MAIL, LETTERS, AND MEMORANDA

As an engineer, you'll often need to compose short memoranda, faxes, e-mail messages, and notes to fellow engineers, supervisors, employees, or customers. Whether sent on paper or electronically, a memo must convey its message clearly and concisely. It must leave no room for ambiguity or misinterpretation, and it must be written in proper grammatical style. When you engage in a face-to-face conversation, you can modify your communication approach on the spot, depending on the person's reaction. But you will not have the benefit of witnessing the reaction of a person who has received a memo. A written memo must be carefully crafted to elicit the desired reaction on the first reading.

### 7.4.1 Writing Electronic Mail Messages

Electronic mail, or e-mail, has become the communication tool of choice for countless individuals worldwide. Engineers certainly are included in this group of e-mail participants. E-mail has become an indispensable part of an engineer's work environment as well as an integral part of the design process. E-mail has many uses, but its most common use is as a replacement for hard-copy memoranda. Before electronic mail, communication between employees was conveyed by distributing paper to each recipient. The method was slow and labor intensive.

Electronic mail has made distribution an instantaneous and incidental event. The act of composing an e-mail message, rightfully so, now takes up the largest fraction a sender's time. In the age of electronic mail, the sender has more time to *think* about what is being written, because almost no time is needed to copy and distribute the document. One of the most common mistakes in e-mail writing, however, is to assume that messages sent electronically do not have to be prepared with the same care as do other written documents. An e-mail message that is sloppy and poorly written will not be taken as seriously as one that is carefully written in a professional manner. The wise writer sets a proper tone by preparing e-mail messages with the same care as paper memos. The recipient of an e-mail message will read it in private, just like a written memo, and may treat it with the same formality as a memo received on paper. In addition, an e-mail memo can be forwarded to large numbers of additional recipients instantly, spawning multiple exposures. Some people print out received e-mail messages, thereby further blurring the distinction between paper and electronic memos.

The formula for writing a competent and effective electronic mail message is easy to learn. A good message should include the following three items:

- A *header* that indicates the recipient, sender, subject, and date.
- A *first sentence* that states the purpose of the message.
- A *body* that delivers the key points of the message.

### 7.4.2 Header

In our information-abundant world, document organization has become a critical component in business, commerce, and all engineering disciplines. A header that announces the *recipient, sender, subject,* and *date* of the e-mail helps both the sender and receiver to categorize the message and properly file it for future reference. Identifying the subject of the message in the header will also set its tone and prepare the reader for the message to follow. Has the memo been sent to a general distribution list? Is it intended for one person's eyes only? Will the message to follow be formal, informal, alarming, or humorous? A properly designed header will set the tone for the message and prepare the reader to receive it. The header of a memo should look something like this:

To: Karin Peterson

From: Frederick Unlu

Subject: Test Data for Motor Evaluation

Date: April 12, 2001

An alternative form suitable for a message sent to a group of people might look like this:

To: Distribution

From: Tina Oulette, Team Leader

RE: Next Team Meeting

April 12, 2001

Note that many electronic mail systems include automatic prompts for header entries, and most all include a date stamp on the message regardless of whether the sender inserts one on his or her own. Nevertheless, it's a good idea to specifically include a header similar to one of the foregoing examples, because the Internet system often adds network routing information between the automated header and the body of the text.

### 7.4.3 First Sentence

A good message will clearly state its purpose in the first, or possibly the second, sentence. A sentence that begins with, "I am writing this message to inform you . . ." will help the reader unambiguously understand why you have written the message. Stating the purpose of the message right at the beginning will help formulate its tone and state its objectives. Will your message provide information, request a response, ask for permission, or give instructions? The structure of the first sentence will determine the way in which the body is received. It will also ensure that the reader does not misinterpret your reason for writing.

**EXAMPLE 7.2: THE TRIP REQUEST**

As an example of the power of the first sentence, consider the following memo that was sent by an employee of an engineering company. The engineer was writing to his boss to ask for permission to attend a conference of the American Society of Mechanical Engineers.

To: Roscoe Varquin

From: Harry Coates

Subject: Upcoming ASME Conference

Date: January 14, 2001

Roscoe,

As you may know, the American Society of Mechanical Engineers is holding a conference on lightweight composites in Dayton, Ohio, at the end of June. I think that someone from our company should attend this meeting. Over the past several years, composites have shown promise as a viable alternative to steel or aluminum. They combine the strength of the former with the light weight of the latter. It's time that we learned more about these important materials.

The conference will be held at the Dayton Buena Vista. I've spoken with our travel agent and found that flights will cost about $400 for a round trip. Hotel will be $82 per day, and the conference registration fee is $180. Please let me know what you think.

*Critique*   Harry began his message by explaining the upcoming conference and proceeded to present his data about how much it would cost him to make the trip. Despite what seemed like a clear explanation, Harry never got to go to the conference. Although Roscoe was convinced of its value to the company, Harry had presented his data but never stated explicitly in his first sentence that it was *he* (Harry) who wanted to attend the conference. The message brought the matter to Roscoe's attention, and it was Roscoe himself who wound up going to the conference! The confusion resulted because Harry failed to clearly state the purpose of his memo. His e-mail message would have been *much* more effective had it begun with the simple sentence, "I would like permission to attend the upcoming ASME conference in June." Harry learned, much after the fact, that it's important to state the purpose of a memo in its first sentence. Had he talked to Roscoe in person, he might have detected the confusion by witnessing his boss's reaction on the spot, but his chosen method of communication omitted the link of personal contact.

## 7.4.4   Body

When composing the body of an e-mail message, follow basic rules of style and grammar. Each idea or concept should have its own paragraph, and paragraphs should never consist of a single sentence only. Each paragraph should follow logically from its predecessor so that the flow of ideas makes sense to the reader. When you compose a message, think about the sequence of ideas so that your message has structure and flows logically. Use the full-screen edit feature of your software, rather than the one-line-at-a-time mode available on some e-mail systems. In this way, you will be able to go back and revise or restructure the body of the message before sending it.

**EXAMPLE 7.3: VISIT TO THE CUSTOMER'S WORKPLACE**

One quality of a good e-mail message is its ability to convey correctly all the subtleties of a given subject. Without personal contact to enhance communication, an e-mail message must be precise and readily comprehended on the first reading. Consider the following not-so-well written message that describes the visit of a software engineer to the client's workplace. Try to identify the deficiencies in the memo. Its subject concerns a program called the Universal Information System. The task involved rewriting an original UNIX version of the program to run under Microsoft Windows98. The software program is used by the client to keep track of customer charge card records. The software engineer in charge was writing a memo to her boss summarizing the results of a recent exploratory visit to the client.

To: Roscoe Varquin
From: L. Berkin
Subject: Universal Information System
February 26, 2001

On the 25th of February, my group met with the customer. We were allowed to use a user's ID in order to bring up his account record on the UIS. The customer crossed out the user's name on the sheet containing his information, but it turned out that the UIS listed his account with both his name and social security number. My group made a record of how the system was set up in order for us to have the same heading in our program.

I learned some of the commands that are used on the UIS Galaxy system. The command TR14 finds the user's in-question account and displays it on the monitor. The TR33 command displays all transactions which the user made since the specified date. Last, but not least, the TR35 command displays all payments made since the given date. It turned out that this command was a recent one designed and implemented by one of our fellow employees who recently left the company. In our program, we will assume that this command will be used by the customer, even though there is doubt that no one would keep it up to date except for the fellow employee who's stay there is not guaranteed.

Meeting with the customer at her office made it clearer what they expect of the product. They want the product easy to use. They want the account summary sheet to highlight all the transactions not yet paid for. In all, I learned several commands and got some exposure to the UIS system. This was a profitable meeting.

*Critique.* Despite its poor grammar and flow of ideas, the body of the message does contain the following key points:

- An opportunity to see a demonstration of the UIS system took place.

- The customer erroneously allowed us to see the identity of a sample account holder.

- A program within the UIS system is called "Galaxy."

- The UIS system includes several commands:

  - TR14: Customer account,

  - TR33: Transactions display, and

  - TR35: Payment history.

- The TR35 command was designed by a previous employee of the company.

- The meeting was useful.

- The customer wants unpaid transactions highlighted.

Regardless of its content, however, the body is deficient, because the order in which the information is presented is entirely random. The message lacks logical progression, contains no preamble, and includes no added comments to aid in interpretation. Each fact is presented with no accompanying explanation of how it fits into the overall context of the memo. Additionally, the information contained in the first paragraph is irrelevant to the writer's boss who would not care that the customer made a mistake while attempting to access the system. The writer seems to have had a list of facts in her head and proceeded to regurgitate them on paper in whatever order they came to mind. The message also omits critical information. The author of the memo does not mention the identity of the customer or who the person was that gave the tour and demonstration. There is no mention of what the letters "UIS" stand for, and the term "Galaxy" is used without explanation. It's highly likely that this memo underwent very little onscreen editing or rewriting and probably was poorly received.

Now consider the following version of the message, which was written and revised using proper grammar, construction, and form prior to being sent.

To: Roscoe Varquin
From: L. Berkin
Subject: Summary of customer meeting at Boulton Industries
February 26, 2001

Roscoe,

This memo summarizes what we learned during our visit to Boulton Industries on February 25, 2001. Our project team met with Boulton's representative, Ms. Connie Donaldson, at their Weston office. The meeting provided us with an overview of the customer's existing Universal Information System (UIS) and introduced us to the procedures commonly used in Ms. Donaldson's office. We worked with her on the UIS system using actual account records.

The UIS system contains an application program called Galaxy that allows the customer to view account records. The record of an account holder is accessed by entering an ID number into the Galaxy system at the menu prompt. The account data can then be viewed in a variety of formats useful to the customer.

Several user programs, called Transaction (TR) screens, reside within Galaxy and help the user to display account information. These TR screens are used routinely by Boulton employees for processing monthly bills and servicing customer call-in inquiries. During our meeting, we recorded the display headings of each important TR screen so that we can create the same headings on our PC-based Windows98 version of the UIS system.

As part of our tour, Ms. Donaldson demonstrated some of the commonly used TR screens. One command, for example, is called TR14. It retrieves a specified account record and displays it as a standard mailing form on the monitor. A second command, called TR33, displays all transactions that have occurred since a date specified by the user. A third command, called TR35, displays the history of all payments made since the user-specified date. This last command was written previously for Boulton by one of our employees. When writing our new version of the program, we should assume that this command again will be used by Boulton to process records, and we should plan to integrate the feature into the Windows version of the program.

I feel that we learned enough from our meeting to enable us to proceed with the Boulton UIS project. If necessary, we can return to Ms. Donaldson's office at a later date to obtain more information.

Sincerely yours,

Laura Berkin
Software Division

## 7.4.5   Writing Formal Memos and Letters

The informality of an e-mail message in not appropriate for all communication. At times, the added formality of a written, hard-copy letter is preferable. Formality suggests importance. Applying for a job or sending a follow-up thank-you letter, for example, are situations that call for a formal written letter. Likewise, if your information has archival quality, or if your message may have legal implications, then you should send your memo in paper format. Besides carrying more social weight than an e-mail message, a paper letter can bear a binding signature.

The rules for composing and sending a written letter are almost identical to those for sending electronic mail. One key difference is that a formal letter normally does not contain a To and From header, but instead it begins with the recipient's address and a formal salutation. A letter should be well presented on good paper or letterhead and be printed in an attractive format. The following example illustrates some of the finer points of writing an effective formal letter. The first version shows the letter as originally written, and the second shows the results of a much needed revision.

**EXAMPLE 7.4: SUMMARY OF TESTS RESULTS**

The following letter was written by an engineer who wished to summarize the results of mechanical loading tests for a client. The letter is based on the following points, listed here in the order in which the author wrote them down in his logbook:

- Initial loading tests are completed.
- Test samples were composites of carbon and epoxy.
- Control samples were steel.
- Same shape was chosen for each; steel was machined, composites were molded.
- Samples were composites and steel.
- Initial difficulties with fitting samples into test machine.
- Made a holding jig to allow testing.
- We should go with composites at slightly increased diameter.
- Numerical data from test results:

| DIAMETER | COMPOSITE | STEEL |
| --- | --- | --- |
| 0.25 in | 245 lbf | 321 lbf |
| 0.375 in | 1644 lbf | 1790 lbf |
| 0.50 in | 3021 lbf | 3229 lbf |

The letter as written by the engineer reads as follows:

Apex Systems
Structural Testing Laboratory
730 Commonwealth Ave., West Roxbury, MA 02132

Helen Brickland
Access Engineering
44 Cummington St.
Boston, MA 02215
January 18, 2001

Dear Ms. Brickland:
We've completed the initial loading tests on the samples made from composites and steel. The samples had the same shape, and the steel was machined and the composites molded. We had some initial difficulties fitting the samples into the test machine but finally made a holding jig to allow testing. I think that we should go with composites at slightly increased diameter.

Here are the pieces of data:

- 0.25 in. diameter: composite 245 lbf, steel 321 lbf
- 0.375 in. diameter: composite 1644 lbf, steel 1790 lbf
- 0.50 in. diameter: composite 3021 lbf, steel 3229 lbf

Sincerely yours,

Ed Garber
Mechanical Engineer

*Critique*    This letter is correctly formed and includes all needed information, but the writer would have done a better job of stating his objectives if the first sentence had cited the real purpose of the letter, which was to make a recommendation to Ms. Brickland that composites be used instead of steel. A better opening might have been the following:

Dear Ms. Brickland:

For the past month, Apex Systems has been performing loadings tests on samples of composites and steel for Access Engineering. Based on our test results, I'd like to recommend that we choose composites for this project.

The original letter has other problems beyond those of its first sentence. Its body, for example, is completely disorganized. Although the author has followed almost verbatim the ordering of ideas as recorded in his logbook, those ideas were not particularly well ordered to begin with. The letter reads disjointedly, as if it's been poorly edited. The sentences are choppy and do not flow from one to another. The numerical data should have been presented in more concise, tabular form. Finally, the author failed to describe the purpose of the tests, didn't go into detail about how they were performed, and didn't describe the test samples other than to say that they were a mix of composites and steel. He should have provided the dimensions of the samples, the composition mix of the composites, and how many of each type of sample were tested.

A revised version of the letter that corrects for these deficiencies might look something like the following:

Apex Systems
Structural Testing Laboratory
730 Commonwealth Ave., West Roxbury, MA 02132

Helen Brickland
Access Engineering
44 Cummington St.
Boston, MA 02215
January 18, 2001

Dear Ms. Brickland:

The mechanical group working on the Delta vehicle project with Access Engineering has just completed tests on samples of the composite and steel materials that we are evaluating for the main structural members of the frame. Based on our test results, our group recommends that we choose composites of slightly enlarged diameter for the structural materials. The details of the tests are described below.

Samples of machined steel and molded composites were fabricated in the shape of standard tensile test specimen bars having a variety of diameters in the range 0.25 in. to 0.625 in. (ASME specification 246). These samples were stressed to the breaking point in our lab's Instron test machine. Although we had some initial difficulties in fitting the samples of diameter other than 0.5 in. into the test machine, we were able to make a holding jig to accommodate testing of all samples. The numerical data from our tests results are summarized in the following table:

| BREAKING FORCE UNDER TENSION | | | | |
|---|---|---|---|---|
| Diameter (in): | 0.250 | 0.375 | 0.500 | 0.625 |
| Composite (lbf): | 845 | 1644 | 3021 | 4229 |
| Steel: | 1421 | 2790 | 4310 | 5541 |

As you can see, our minimum targeted breaking strength of 4000 lbf can be met with either a composite rod of 5/8-in (0.625″) diameter or a steel rod of 1/2-in (0.500″) diameter. Given the much lighter weight of the composite material, however, its strength-to-weight ratio is much higher than that of steel. I recommend that your company go with composites for this project.

Sincerely yours,

Ed Garber

Mechanical Engineer

This second version is much improved over the original. The author has articulated the purpose of the letter in the second sentence, rather than in the first, but it works well here because the letter still reads and flows nicely. The first sentence serves as a preamble to the real purpose of the letter which is revealed in the sentence that begins, "Based on our test results . . ."

The data are also well presented in the table contained within the letter. Instead of putting the units next to each entry, Ed has written them just once next to the category titles on the left-hand side. This format makes for a much neater display of data. Ed also has elaborated on the details of the tests. When Ms. Brickland reads the letter, the context in which the tests were taken will immediately be clear. This point is an important one. Ed might falsely conclude that Ms. Brickland will understand its context, because his primary work assignment over the past month has been the taking of data and the testing of the materials. But Helen, as a manager, is likely to be juggling dozens of projects and details, and she will welcome a letter that first refreshes her memory about its context and background. In addition, she may find herself reading the letter at some future time when the background details of the test procedures have faded from memory. Or she might forward the letter to someone else who is unfamiliar with the details that Ed has correctly provided.

**PROFESSIONAL SUCCESS: PREPARING A PRESENTATION FOR A NONTECHNICAL AUDIENCE**

At times an engineer must make a presentation to a nontechnical audience. You also may encounter this challenge while you are a student. Perhaps you'll be invited to speak at your old high school. Maybe you'll be asked to explain the activities of a research laboratory to visiting parents and prospective students. Or maybe you'll receive a class assignment to prepare a talk on a technical subject for a general audience. In these and similar situations, the following few basic principles can guide you:

- Assume the audience knows nothing about your topic.
- Explain background material without using jargon words. (You'd be surprised at how many seemingly common words are really part of the engineer's private lexicon.)
- Start at the beginning to provide the big picture.
- Pretend that you're speaking to a group of fourth graders.
- Never show equations to a nontechnical audience. (You'll probably pique their math anxiety.)

## 7.5 WRITING TECHNICAL REPORTS, PROPOSALS, AND JOURNAL ARTICLES

Technical reports, proposals, and journal articles are the engineer's equivalent of term papers. As an engineer, you are likely to face the task of writing one of these documents at some point in your career. Unlike short e-mail messages, memos, and letters, which are usually only a few paragraphs long, longer technical documents require considerable thought and preparation and usually cannot be finished in one sitting. The sections that follow highlight some of the key elements of these longer documents.

### 7.5.1 Technical Report

A technical report is used to convey important findings or test results to a controlled audience. Technical reports seldom undergo peer review, and distribution of the report is done at the discretion of the author or employer.

The typical technical report is between two and twenty pages long. In content and form, it's not unlike a lab report you might prepare in some of your college courses. Although the format may vary somewhat, most technical reports contain the following elements: Introduction (or Background), Experimental Setup (if applicable), Theory, Data, Analysis, and Conclusion.

The introduction serves as a preamble to the document and states its reason for having been written. Unlike the first sentence of a simple memo, the introduction of a technical report can occupy a paragraph, a page, or sometimes many pages. A person pressed for time will merely skim the introduction, so it, along with the conclusion, should provide a self-contained overview of the entire report.

If the document describes the results of an experiment, a section should be included that describes the physical setup. This section should provide enough detail such that a reasonably competent person could completely reconstruct the experimental apparatus and obtain similar results. It should describe instruments, apparatus, mechanical techniques, dimensions, and other key parameters.

The data section includes the results of any experiments or tests that were performed. It should explain why each set of data is presented, how it was obtained, and what bearing it has on the main purpose of the document. A report or journal paper is likely to be used later as a reference source, so it's important to present data completely and accurately. The presentation should be easily digested by someone not intimately familiar with the details of the project.

The analysis section is where the data are evaluated, interpreted, and used to support any claims made in the report. Mathematical calculations belong in this section, as do plots and charts derived from the data. In some cases, particularly in reports that deal with design work, the analysis and data sections appear in reverse order. First the analysis of the device is presented, followed by data on tests that show whether the device meets the predictions of the analysis.

Finally, the conclusion is used to summarize the claims, results, and observations included in the report. Some individuals may not have time to read the whole report, but need to be familiar with its content. The conclusion section should be written to serve the needs of a person who only has time to browse or skim the report. The conclusion should be a stand-alone section that summarizes all the key points of the report.

### 7.5.2 Journal Paper

Journal papers provide a way in which engineers disseminate information of interest to other engineers. They also provide a public forum in which engineers and scientists can announce "first to invent" status of a new technology or discovery. Most journal papers, particularly those sponsored by professional organizations, must undergo peer review before publication. This procedure helps insure their quality and accuracy. Although the standard format—introduction, theory, experiment, data, analysis, and conclusion—is appropriate for journal papers, many publications specify their own format in which a journal paper must be submitted.

### 7.5.3 Proposal

A proposal differs from a report or journal paper in that its primary objective is usually to secure *money*. A proposal attempts to convince a client or funding source that your organization can best handle a research or design job, or perhaps that your product will be best for a particular application. In addition to the various sections of a technical report, a proposal often includes additional sections on objective, budget, company background, and personnel.

## 7.6    PREPARING AN INSTRUCTION MANUAL

One of the most common documents written by engineers is the instruction manual. An instruction manual introduces the user to your product and provides information regarding its setup, operation, and use. A well-written instruction manual also includes sections on safety information, troubleshooting, repair, and theory of operation, if applicable. While not all engineered devices require an instruction manual (the operation of a snow shovel, for example, should be self-explanatory), those that involve detailed operating procedures should be accompanied by instructions. Indeed, user perception of many products is derived directly from the quality of the instruction manual.

The sections of a typical instruction manual are outlined below. If the manual is long, a table of contents with page numbers should be included. Obviously, the need for each of the sections suggested below will depend on the specific product described by the manual.

### 7.6.1    Introduction

The introduction should provide an overview of the product. It should explain the purpose of the product, its usefulness to the user, its special features, and proper use of the manual itself. Various titles for the introductory section include "Getting Started," "Welcome to Product X," "Before Using Your Device," etc.

### 7.6.2    Setup

The setup section should outline the procedures that the user must follow before the product is ready for use. This section should appear at the front of the manual, where it will be easy to find. It should guide the user through the setup procedure step by step, and it should use illustrations liberally.

### 7.6.3    Operation

The operation section of the instruction manual is its most important part. A first-time user will use this section to learn how to operate the product and will refer to it thereafter to clarify points of operation. Because of its likely use as a reference source, the operation section should be carefully laid out and organized so that the user can extract information without having to read the entire manual from the beginning.

### 7.6.4    Safety

Safety is a very important aspect of engineering design, and any such information relevant to the well-being of the user must be included in the instruction manual. If the product has dangerous moving parts or high voltages, includes safety panels or guards that must not be removed, or has the potential to emit flying objects or capture loose clothing, then appropriate warnings should be included. In our overly litigious society, where personal injury lawyers advertise freely on the radio and TV, safety warnings have become pervasive. Some safety warnings may seem overly cautious or even ridiculous (e.g., don't put your fingers in the moving blades or you might get hurt), but safety warnings have become a necessary part of engineering.

### 7.6.5    Troubleshooting

Expect your product to break down, regardless of how well it is made. If it should happen not to malfunction over its lifetime, consider yourself lucky. Designs do fail, and the troubleshooting section should guide the user through a simple set of tests that can help identify the source of any malfunction and get the system running again. The section should outline simple repairs that the user can try before taking the more drastic step of returning the product. If appropriate, include a section that explains how to get in touch with the manufacturer (you) in case difficulties cannot be resolved by the user.

Never assume that the user will understand something that is not clearly specified. The following troubleshooting entries may seem silly, but many instruction manuals contain similar versions:

*Symptom*: No lights or displays of any kind are lit; unit appears to be dead.
*Possible Cause*: Unit is unplugged or has a blown fuse.
*Remedy*: Plug cord into proper outlet. Replace fuse.

*Symptom*: No sound is coming from the speaker.
*Possible Cause*: Volume control is turned all the way down.
*Remedy*: Turn volume knob clockwise.

*Symptom*: Drive shaft does not turn.
*Possible Cause*: Clutch is not engaged.
*Remedy*: Engage clutch by moving lever to the "on" position.

*Symptom*: No flame emits from burner.
*Possible Cause*: Pilot light is extinguished.
*Remedy*: Relight pilot. (See Section 2.1: Lighting the Pilot Light.)

*Symptom*: Pilot light cannot be lit.
*Possible Cause*: No gas supply.
*Remedy*: Open main gas valve; replace propane tank.

### 7.6.6   Appendices

Information likely to be of interest only to a special readership should be included in appendices. Examples include circuit schematics, exploded assembly diagrams, theory of operation, and lists of part numbers.

### 7.6.7   Repetition

One of the subtle features of a good instruction manual is its ability to engage the reader regardless of where he or she begins reading it. This attribute can be achieved by repeating informative details at several junction points throughout the document. As you write the manual, imagine the viewpoint of a reader who has begun to read it somewhere in the middle of the document. Restate key information at topical transitions and major section headings, rather than referring to previous sections. Do not assume that the reader remembers details covered previously if they are relevant to the present section.

**EXAMPLE 7.5: THE ATM SIMULATOR**

The following example contains (slightly edited) excerpts from an instruction manual written by senior project students in the Department of Electrical and Computer Engineering at Boston University.[3] It includes many of the elements mentioned above and illustrates the overall features of a good instruction manual. The manual explains the operation of a simulator for a bank automatic teller machine designed to teach banking procedures to elementary school and special needs students.

**Automated Teller Machine Simulator**
**Instruction Manual**
**Terrier Technologies of Boston**

*WELCOME TO THE TTB ATM SIMULATOR*
The Terrier Technologies of Boston Automated Teller Machine simulator has been designed to help you teach students the important aspects of banking skills. The unit has been designed for simplicity and ease of use. It should provide you with many years of trouble-free operation.

*HOW TO USE THIS MANUAL*
The TTB ATM users' manual is divided into five sections and an appendix. The first two sections provide a general overview of setup and system start-up, respectively. The third

---

[3] "G. DeBernardi, R. DeMayo, M. Givens, M. Magne, E. McMorrow, and S. Tansi, *Automated Teller Machine Simulator Instruction Manual*. Terriers Technologies of Boston, 1992.

section explores the inside of the ATM simulator and discusses its various modules. The remaining two sections deal with care and troubleshooting of the equipment. The appendix contains wiring diagrams and computer software codes.

## OVERVIEW OF OPERATION

The TTB ATM simulator is a self-contained product. Each of its modules simulates the operation of a real bank ATM machine. Account information is stored in the computer connected to the ATM panel. A banking session begins by prompting the user to insert a card into the card slot. Once the card has been properly inserted, the user is asked for a password to be entered via the keypad. After entering the correct password, the user is asked to choose from the following list of transactions:

1.   Deposit

2.   Withdrawal

3.   Fast Cash

4.   Account Balances

Once the type of transaction has been chosen, the simulator asks the user to choose between a savings account and a checking account. The user can withdraw facsimile money from the cash dispenser or place an envelope in the deposit slot. Immediately following the transaction, a receipt of the session is printed out and the user's account is updated inside the computer.

The system operator has additional choices not shown on the main user menu. These additional choices include: `Modify Parameters`, `Troubleshooting`, and `Print Account Information`. The `Modify Parameters` command enables the system operator to update user accounts. The `Troubleshooting` command tests the individual modules of the ATM simulator. `Print Account Information` makes a printout of all user names and account balances. The ATM simulator is also equipped with two flip-up panels. The top panel exposes the main keyboard used by the system operator to initiate and update accounts. The bottom panel provides access for paper replacement for the receipt printer and provides access to the cash dispenser.

## INITIAL START-UP (SYSTEM OPERATOR)

The power switch is located on the rear of the unit. Plug in the power cord and move the power switch to the ON position. You should hear a whirring sound as the internal disk drive is activated. To begin the session for entering or updating user accounts, first access the keyboard by raising its hinged cover. The screen display will give instructions on how to proceed.

## SETTING UP OR CHANGING AN ACCOUNT

To set up a new account or change a previous account, access `Modify Parameters` from the `Main Menu`. Once this selection has been made, the following set of selections will appear on the screen:

1.   ID Number

2.   Password Savings

3.   Checking

4.   Account Balances

Choose your selection by pressing the appropriate number on the numeric keypad. Using the arrow keys, move the cursor to the entry for the user account to be edited. The following set of commands can be used to enter and edit field data:

`Enter`: Access the field that needs to be added or edited.

`Ins`: Add information to a user field.

`Del`: Delete information from a user field.

`Esc`: Return to the Main Menu.

All fields must be filled if the system is to access each user account properly. Once all account information has been entered, a full printout can be made by choosing `Print Account Information` from the `Main Menu`.

### BEGINNING A BANKING SESSION

When all user accounts have been entered, the unit is ready for simulated banking sessions. From the `Main Menu`, select `Begin Session` to begin a simulation. The screens will guide the user through the entire process in a manner very similar to a real automated teller machine.

### INSIDE THE ATM SIMULATOR

An IBM 8088 computer is used to control all aspects of operation for the simulator. The system operator controls the account balances and all other program functions via the keyboard. The display consists of a 13-inch monochrome monitor. The screen guides the user through the entire process and provides help whenever necessary. Display screens are similar to those found on real ATM bank machines. Additional screens take the system operator through the account information sequences.

### KEYPAD

The TTB ATM keypad is identical to that found on real Diebold ATM machines. It consists of fifteen keys (11 blue and 4 white). The 0 through 9 blue digits and decimal-point keys are used for selecting dollar amounts, and the four white keys, labeled A, B, C, and D, are used for making transaction decisions. Pressing the CANCEL key will terminate the session at any given time.

### CARD SLOT

After the system operator has set up user accounts, the simulator will wait for the insertion of a bank card. The card must be placed into the machine in the proper direction shown on the diagram on the ATM front panel. The card will be pulled into the slot by the TTB ATM and will remain in place unless the card is inserted in the wrong direction, the transaction is terminated by the user, or the system is shut down or loses power.

### MONEY DISPENSER

The user may ask the ATM for a withdrawal in increments of ten dollars. If the user asks for cash in other increments, the simulator will inform the user that the transaction is not allowed and will suggest trying again. Once an amount has been specified for withdrawal, the cash dispenser will drop facsimile ten-dollar bills into the money bin. The system operator can reload the machine with money when needed.

### RELOADING

When bills need to be reloaded, the operator should access the cash dispenser from the back of the TTB ATM by pushing down on the springs and placing a neatly packed stack of facsimile bills into the unit. Be sure that the bills can roll out freely by manually turning the dispensing wheels until the bills can move easily.

### DEPOSIT SLOT

As in most ATM machines, money or checks can be deposited by placing a deposit envelope into the deposit slot. After the simulator has informed the user that it is ready to receive a deposit, it awaits the insertion of an envelope. The deposit slot module will pull the envelope inside the unit. Deposit envelopes may be collected later by the system operator from the deposit bin. An indicator light located on the back panel of the TTB ATM will inform the system operator if a deposit envelope has been received. Once the envelope has been securely received by the ATM, the user's account will be updated automatically.

### PRINTER

After a transaction has been completed, the simulator will produce a copy of the transaction and balance information. The printer is mounted on the side of the unit for easy access. The paper is the same type used in adding machines. Its roll should be placed on the bar so that it can be fed into the printer. When the paper roll runs out, it can be easily replaced by sliding a new roll on to the bar. The printer is connected to the computer via a standard Centronics printer interface cable.

### EQUIPMENT MAINTENANCE

The main unit may be cleaned with a damp (not wet), lint-free cloth. Aerosol sprays or other cleaning solvents are not recommended for cleaning. If the screen gets dirty, use a clean cloth or paper towel to wipe the screen.

The unit contains no user serviceable internal parts. Repairs should be performed only by a qualified TTB technician. Moving or tampering with any of the components of the TTB ATM may adversely affect the operation of the overall system.

### SAFETY PRECAUTIONS

Even though the TTB ATM unit is designed to be as safe as possible, a few circumstances can lead to hazard conditions. For your own safety, and that of your equipment, always take the following precautions. Disconnect the power plug if

- the power cord or plug becomes damaged, or
- any piece of clothing gets caught in the bank card, printer, envelope, or money dispensing slots, or
- any liquid is spilled on the unit, or
- the ATM simulator is dropped.

### TROUBLESHOOTING

*Problem*: No lights or displays of any kind are lit; unit appears to be dead.
*Possible Cause*: Unit unplugged or has a blown fuse.
*Problem*: Keys are stuck.
*Possible Cause*: Keys may begin to stick due to temperature or excessive use. Try to loosen stuck keys by gently pulling them up. WARNING: Do not pull too hard or the keys may break off.

*Problem*: No response from keypad.

*Possible Cause*: The TTB ATM may not be in transaction simulation mode.

*Problem*: Envelope is not being pulled into the deposit slot.

*Possible Cause*: The wheel of the drive motor may be stuck. Pull up on the traction wheel or remove the envelope and push it manually through the slot.

*Problem*: Bills are not being dispensed from the machine.

*Possible Cause*: The money dispenser may be empty.

*SERVICE AND SUPPORT*

Service for the ATM simulator is available through our local field service network. Please contact:

Terrier Technologies of Boston
ECE Department, Boston University
8 Saint Mary's St.
Boston, MA 02215
617-353-9052

## 7.7 PRODUCING GOOD TECHNICAL DOCUMENTS: A STRATEGY

With the possible exceptions of the short memo and e-mail message, writing good documents takes preparation, time, and effort. Whether your writing task consists of an instruction manual, technical report, journal paper, annual summary, or an entire book, it will stand a better chance of accomplishing its objectives if it is well written. As with any other skill, learning to write well requires practice, patience, and attention to detail. In this section, we review several time-tested techniques for improving your writing abilities. Although different writers develop individual styles, most follow the same basic rules outlined below.

### 7.7.1 Plan the Writing Task

Before sitting down to write, gather all pertinent information. Assemble the results of design calculations, tests, experiments, user specifications, and all other available material. If it's pertinent to the writing task, make sure your engineering logbook is by your side. Gather reference citations, figures, and graphics, if applicable. Have everything at your disposal before you begin the writing task.

### 7.7.2 Find a Place to Work

One of the most important lessons to learn about writing long documents is that you must devote an uninterrupted block of time to the job. It's simply not possible to write well if you are distracted by telephones, e-mail, television, or people coming to talk to you. Writing a complex document takes a long time, sometimes hours or days. When you face a writing task, persistent concentration over an extended period of time will help get you into a creative writing mode. Your mind must sort ideas, arrange their flow, and commit them to well-written and enticing prose. Choosing precise words is an art form similar to painting, sculpting, or composing music. Your writing will flow more smoothly if you find a secluded spot where you'll have absolutely no interruptions. The telephone, your e-mail terminal, or other people stopping by will almost certainly break your concentration during writing and interrupt the creative process. In an ideal world, writers would be able to post "Writing in Progress: Do Not Disturb" signs over office doorways or cubicle portholes. In the real world, such sequestered spots are not always available. Go wherever you can find privacy. A library, cafeteria (during off hours), lab bench, or even the corner donut shop provide mental seclusion, despite their public nature, because they provide uninterrupted time for writing.

### 7.7.3   Define the Reader

Decide who will be reading your document. Some readers will know more about your subject than you do. Others will know nothing at all. It's important to know the technical level of your reader so that you can set the *tone* of the document. Suppose, for example, that you are reporting on loading tests for a group of nonengineers. In such a case, you probably wouldn't include material on spring constants, test methods, or Young's modulus of elasticity. If the report were for engineering students or professors, you might want to include these items.

Regardless of the technical level of your readers, you should decide how much detail the reader will need about the topic. Also be aware of what the reader will do with the document. Will it be redistributed? Will someone else read it? Answering these questions will help you set the tone of the document.

### 7.7.4   Make Notes

Professional writers always seem to get their documents to read just right. You, too, can produce well-organized, easy-to-read documents by mastering one valuable method used by the professionals. Before you begin writing the actual document, make random, stream-of consciousness notes—one line reminders—of *anything* that might need to go into the document. At this stage, give no particular attention to order or emphasis. Include the obvious essentials, as well as the possibly needless trivia. Many writers find it more effective to perform this step with traditional paper and pencil rather than on a computer, because the act of keyboard typing is known to occupy a sizable fraction of brain activity, leaving less mental power for creative and organizational activities. Regardless of which method you choose for recording your ideas, the key at this stage is not to worry about the order in which you write things down on your list. Commit all your ideas to paper now. You'll scrutinize them at a later step in the writing process.

### 7.7.5   Create Topic Headings

Your next step should be to form the overall structure of the document. To accomplish this task, you should write down the topic headings that will need to go into the finished work. Again, you should write these items down in random order, paying no attention at this stage to how they will be structured. Each of these topic headings eventually will become a paragraph or sequence of paragraphs in the finished document. When you're done with your list, examine each topic heading to see if additional headings come to mind. Delete irrelevant headings and group remaining headings into the main topic areas of the document.

When your list of topic headings is complete, arrange them in a suitable order. Decide which order of presentation is the most interesting, logical, and easiest to understand. It's at this point that the main structural framework of the document begins to take shape.

### 7.7.6   Take a Break

Before you begin to write the actual document, take a break. Clear your mind before beginning the writing process.

### 7.7.7   Write the First Draft

If you've done a good job of preparing your list of one-line notes, you'll be ready to begin the writing process. Find an interruption-free spot to work, and start to write.

Don't worry about writing in perfect form at this stage. Expect to revise your document many times before it's completed. The important thing during the first draft stage is to get your words down on paper. Use a word processor or write by hand and type later, whichever method suits you. As previously mentioned, using a keyboard detracts from the creative energy of writing for some people, so don't feel that you must compose the first draft on a word processor. A word processor is an indispensable tool for writing, of course, but it's main advantage comes in the revision process. On the other hand, if you *can* learn to compose the first draft directly on a word processor, you'll save a lot of time otherwise spent in transcribing handwritten pages.

At this point in the writing process, don't be too concerned about spelling or exact phrasing. These aspects of the document will be corrected and modified later, in the revision phase.

If the work is short, write the first draft in one work session. If the document is long, divide the work into medium-length work sessions. At the end of each work session, rapidly scan the draft and make only *obvious* changes. Do not do major revisions at this time. When the draft is finished, again take a break to clear your mind. If time permits, set aside the document for another day so that you can approach it with a fresh perspective.

### 7.7.8 Read the Draft

After your break, when you are no longer intently focused on the document, reread it as if you are seeing it for the first time. Check your writing style for clarity. Are there vague, confusing, or ambiguous passages? Are the sentences in the correct order? Is there a logical flow within each paragraph and between successive paragraphs? Check for correct tone. Is the writing style suitable for both the subject matter and the reader?

Try not to read the draft solely from your computer screen. A document always looks different when it's been printed out on paper, because the reader is able to absorb more text at once and get a better feeling for how the entire document is structured.

### 7.7.9 Revise the Draft

A document seldom is ready for distribution after the writing of its first draft. After you write the first draft, take the time to review your work. Revise words, reword sentences, rearrange paragraphs, and reorganize sections to further refine and clarify meaning. As you revise, mercilessly slash unnecessary words and sentences. Weigh each word and phrase, and keep only those words and phrases that carry important meaning. Technical writing should be direct and to the point. Keep each paragraph relevant. Replace complicated phrases with simple words, and limit superlatives. As you reread your document, give it at least one pass in which you ignore content and look at words, sentences, and paragraphs in their grammatical context only. Remove "fat," unnecessary words and details that have low information content. You should also recheck factual statements, formulas, numbers, and calculations for accuracy. Be careful to proofread material cited from other documents. Almost all word processors include a spelling checker. Get into the habit of using one before sending out a document of any kind.

### 7.7.10 Revise, Revise, and Revise Again

After you've made your first revision, revise, revise, and revise again. Most good writers devote three or more, and sometimes dozens, of rewrites to get a document to read just right. Each chapter of the book you are reading now was revised at least six times before being sent to the editor for publication.

### 7.7.11 Review the Final Draft

When you feel that your document is finished, put it aside and come back to it, preferably on another day when you will not be prejudiced by the intensity of the writing process. As you read your document, evaluate it as an outside reader would. Keep an open mind and ask yourself the question, "How would *I* react to what I have written? Will it produce the intended reaction or response from the reader?" If the answer is "yes," your document is ready for the outside world.

### 7.7.12 Common Writing Errors

Errors in usage and grammar are common in work prepared by student writers. Learning to write well takes practice, discipline, and the careful advice of a good teacher. Despite this observation, it is possible to learn some elements of good writing from a text such as this one. In particular, understanding and avoiding common writing errors will help you immensely as you try to develop good writing habits. The writing errors listed in the following sections are typical of those found in written assignments submitted by engineering students. Review them so that you can avoid making similar mistakes in your own work. Additional examples of correct and incorrect usages can be found in references such as Strunk and White (1979).[4]

*Parallelism.*    Sentences that include multiple items or ideas should follow parallel construction.

*Correct:* "Our module will provide data communication, consume minimal power, and satisfy the customer's needs." (All three phrases begin with a verb.)

*Incorrect:* "Our module will provide data communication, minimal power will be consumed by it, and it will satisfy the customer's needs." (The three clauses don't have the same construction.)

*Commas.*    Use a comma to separate the second part of a sentence only when the second half could stand on its own as a complete sentence.

*Correct* (do use a comma): "We will supply five commands to the robot, and we will power the robot with batteries." (The second half of the sentence, "We will power the robot with batteries," is a complete sentence.)

*Incorrect* (don't use a comma): "We will supply five commands to the robot, and power it with batteries." (The second half of the sentence, "and power it with batteries," cannot stand on its own as a separate sentence. The comma after the word "robot" should be omitted.)

*Past, Present, and Future Tense.*    Maintain consistent tense (past, present, or future) as your writing progresses, or at least within a given paragraph.

*Correct:* "The routes will be difficult to change once they have been programmed into memory. This drawback also will apply to future versions of the robot." (future, future)

*Incorrect:* "The routes will be difficult to change once they have been programmed into memory. This drawback also applies to future versions of the robot." (future, present)

*Use of the Word "This."*    For clarity the word "this" is best used as an adjective, not a noun or pronoun. It should be accompanied by an object of reference.

---

[4] W. Strunk and E.B. White, *The Elements of Style.* New York: MacMillan, 1979.

*Correct:* "This problem will be solved by designing a new system."
*Incorrect:* "This will be solved by designing a new system." (i.e., this . . . what?)

**Use of the Words "Input" and "Output."** "Input" and "output" are best used as nouns. Their use as verbs is often awkward and unprofessional.
*Correct:* "The input to the mixing circuit consisted of three microphone voltage signals. The output was fed to the amplifier in the form of a voltage summation."
*Incorrect:* "The microphone signals were inputted to the mixer. Their combined sum was outputted to the amplifier."

**Punctuation Around Parentheses.** When words are set aside by parentheses, place periods *before* the trailing parenthesis if the parenthetical thought is a major part of the sentence.
*Correct:* "Our design project was completed on time. (We had been given a week to complete it.)"
*Incorrect:* "Our design project was completed on time. (We had been given a week to complete it)."

**Infinitives ("To" Verbs).** Never split an infinitive. If you use the word "to" followed by a verb, do not put words in between.
*Correct:* "The purpose of this section is also to help you with your homework."
*Incorrect:* "The purpose of this section is to also help you with your homework."

## KEY TERMS

| | | |
|---|---|---|
| E-mail | Memorandum | Presentation |
| Report | Instruction manual | |

## PROBLEMS

1. Write a report that outlines the design process for a hybrid gasoline electric vehicle design competition.

2. The following document outlines the approach to be taken in the design of a system for paging contest participants over a three-minute time interval as part of an engineering design competition. It is an example of very poor writing. Rewrite the proposal, taking into account the writing principles and suggestions outlined in this chapter.

   The three minute pager receiver will be based on a simple bandpass filter that is tuned to a distinct RF band for each receiver. Additionally, each receiver will tune into a general public announcement band which will broadcast voice messages or tones. The cost will be held very small by constructing our own receiver circuits.

   Power consumption is minimized by sleep mode. In sleep mode, the receiver's PA band amplifier will be disconnected from its power source via a relay or power monitor switch. Detection of a wakeup signal on the wakeup band will close the circuit between the PA band's amplifier and the speaker.

   In addition, preliminary cost research shows that a three-minute countdown circuit and LCD screen can be constructed for under nine dollars in quantities of 100. A speaker and blinking LED can also be provided at minimal cost.

   The countdown itself would also be initiated by reception of the wakeup signal. The end of the internal countdown would power down the PA band amplifier, or a second detection on the wakeup band would toggle the power off.

The unit itself could be wearable and styled after a pager or smartcard. We have scheduled a meeting for Tuesday, January 21, at 11 am.

3.   The following memo was written by an engineer responsible for designing a parts counting device. The writing style is very poor. Rewrite the memo using the guidelines discussed in this chapter.

During our first conversation with the customer, we came to an initial design for the project and have scheduled a meeting with the clients on Tuesday. For the design, first thing come across is a detector to physically count the parts falling through the sorting mechanism and we generally prefer to use a photosensor. For the counting mechanism, two methods are proposed and yet remain undecided. One of them is to program a PLA whose reprogramming process could be too complicated for the end user. However, it's advantage is that the design would be simple and cheap. Another approach is to use a microprocessor to do the counting. However, since our team don't have any experience on this subject before, we are still seeking advice and reference. Finally, when the designated no. of parts are counted, the counter will activate a visual and audio signal which prompts the user that parts are ready for packaging. Then the user can put a plastic bag underneath the container and push a button which opens up the bottom of the container.

The above would conclude our initial idea and we will come up with more details and specifications after the meeting.

4.   Write a memo to your fictitious boss asking permission to attend a technical conference.

5.   Write a short instruction manual that explains how to operate your VCR.

6.   Write a memo to all students in your laboratory addressing the importance of safety procedures and protocol.

7.   Write a memo that summarizes the following information relating to data communication protocol:

Data Protocol:
*DCE = Data Communication Equipment (female connector)
Computer, processor, host: receives data, decodes, establishes communications
*DTE = Data Terminal Equipment (male connector)
Terminal, printer, data board - Sends data and displays output
Parallel Data: 8 or 16 bits of data sent simultaneously with a DR (Data Ready) strobe from the DTE to the DCE and a CTS (Clear to Send) signal from the DCE
Serial Data: 1 start bit; 8 data bits; 2 stop bits, 14,400 baud (bits audio); no parity
Synchronous data - a clock line must be established between DCE and DTE
Asynchronous data - relies on nearly precise timing and start/stop bits
RS-232 standard (receive-send asynchronous data)
Positive and negative voltages (MARK = 1 = NEG; ZERO = 0 = POS)
Held in MARK state when not in use
DB-25 Connector: pin 1 - shield
pin 2: transmit data to DCE
pin 3: receive data at DTE
pin 5: clear to send (CTS) from DCE to DTE
pin 7: signal ground
Note: The DTE sends data on pin 2

8.   Write a proposal to your student governing body asking for money to start an on-campus amateur radio club.

9. The following memo was handed in by the employee of a small company specializing in adaptive aides for physically challenged individuals. The memo is not written particularly well. Rewrite the memo using the principles and guidelines outlined in this chapter.

   To: Xebec Management
   From: H. Chew

   This project is to work with a 47-year-old individual who has no speech capability and limited physical abilities. The subject groans and grunts to indicate discomfort, displeasure, requesting, and refusing. During dinner, our customer would like us to provide the subject with a means to indicate "I want more," "I want something else," and "I want a link," etc.

   To solve this problem, we had called the customer for more information, and she would give us a video tape that is talking about the older person. We also had a team meeting to discuss the project. At the end of the meeting, we considered that we would design a box which consists of an interface panel and a data control unit. The interface panel would consist of four 2.5" buttons. Each button represents a prerecorded phrase. The sets of outputs from the buttons will correspond to the mode selected. The data control unit consists of the power supply, speech memory, speech synthesizer, and audio amplifier.

10. The following entries were collected by a design team working on a major software project. These notes are to be used by the team to write a summary report to the project manager indicating the features that the finished product must have. The software is to be a voice synthesizer system that will enable individuals with impaired speech capability and limited motor skills to communicate by way of a simple computer mouse. Using the rough notes provided by the design team, write the finished report to the project manager.

    - Topics covered include the alarm, requests, and greetings portions of the user interface.
    - The requests frame should be configured to allow the user to express common requests.
    - The greetings frame must have the most common greetings and be designed for easy access.
    - All frames should give the user the option to reconfigure them in any order desired.
    - The alarm frame will consist of five buttons for use in an emergency only.
    - Alarm messages include help, pain, fire, police, and ambulance.
    - Requests include drink, food, bathroom, book, pen, television, radio, and music.
    - Greetings include hello, goodbye, good morning, good evening, and good night.

11. Compose the text of an e-mail message that announces a meeting of your design team for a design competition called "Peak Performance."

12. Compose the text of an e-mail message in which you request a meeting with your boss to discuss a possible raise in salary.

13. Compose the text of an e-mail message in which you request a meeting with your professor to discuss a possible change in your course grade.

14. Compose the text of an e-mail message in which you ask a semiconductor manufacturer to send you a free sample of a microprocessor chip.

15. Compose the text of an e-mail message in which you inform a client about your travel plans for an upcoming technical review meeting.

16. Compose the text of an e-mail message in which you provide arrival information for a government contractor who is coming to your laboratory to review your work.

17. Compose the text of an e-mail message in which you ask for volunteers for a committee to review company safety standards.

18. Compose the text of an e-mail message in which you solicit volunteers to participate in the annual company blood drive for a local hospital.

19. Compose the text of an e-mail message in which you ask your boss for permission to attend the annual conference of the Control and Automation Group of the Institute of Electrical and Electronics Engineers (IEEE).

20. Compose the text of an e-mail message in which you reassure a nervous customer that your prototype for a manufacturing system will be shipped on time.

21. Prepare a set of overhead slides in which you outline your design approach for a car design competition called "Peak Performance."

22. Prepare a set of overhead slides in which you describe the results of combustion tests on a new aircraft engine.

23. Prepare a set of overhead slides in which you outline plans for a proposed new light-rail transportation system in a metropolitan area.

24. Prepare a set of overhead slides in which you describe the benefits of a proposed new graphical user interface for a record-keeping system for a real estate company.

25. Prepare a set of overhead slides in which you describe the important features of a professional quality mountain bicycle.

26. Prepare a set of overhead slides in which you report the results of tests on a high-speed data link for transferring cell phone routing information from site to site.

27. Write a letter to the human resources director of Alpha Corporation in which you reply to a classified advertisement seeking software engineers.

28. Write a letter to the human resources director of Beta Corporation in which you reply to a classified advertisement seeking entry-level mechanical design engineers.

29. Write a letter to the human resources director of Gamma Corporation in which you reply to a classified advertisement seeking biomedical engineers for synthetic drug development.

30. Write a letter to the human resources director of Delta Corporation in which you reply to a classified advertisement seeking mechanical engineers to work on developing jet engines.

31. Write a letter to the human resources director of XYZ Corporation in which you reply to a classified advertisement seeking industrial engineers to design manufacturing systems.

32. Write a letter to the human resources director of Omega State Highway Department in which you reply to a classified advertisement seeking civil engineers for highway construction.

33. Write a letter to the graduate admissions director of State University in which you request information about possible financial aid for Master's degree study.

34. Write a letter to the CEO of your company in which you highlight the details of an unethical practice that you have uncovered within the company.

35. Write a letter to the sales manager of your company in which you detail the virtues of the new pencil that your engineering division has designed.

# 8

# The Day of the Peak Performance Design Competition

The day of the Peak Performance Design Competition has arrived, and you are set to participate in the contest. You've been working on a wedge-shaped design, and you hope your slow, but strong, sturdy vehicle will be able to push faster but weaker opponents from the top of the hill. You are optimistic that your design strategy will pay off, but you are wary of several worthy opponents that are testing their cars at the vehicle preparation tables. One such vehicle, shown at the starting line in Figure 8.1, is built from lightweight epoxy board. It has large wheels and is extremely fast.[1] Although this vehicle is flimsy compared to your aluminum-encased wedge, its designers have installed barbed hooks on the leading edge of the chassis. These hooks are meant to dig into the carpet-covered ramp and prevent backward motion when an opposing vehicle pushes from the front. You decide to call this vehicle "Fish Hook." Fortunately, you're not the first to compete against it, so you have ample opportunity to watch Fish Hook engage other vehicles.

## SECTIONS

*    Epilogue: The Day of the Peak Performance Design Competition

## OBJECTIVES

*In this chapter, you will learn about*

*   The outcome of one particular Peak Performance Design Competition

---

[1] The figures of this chapter show actual photos taken at the Peak Performance Design Competition at Boston University. The stories and scenarios presented are entirely fictitious. For more information about getting started with your own Peak-Performance design competition, contact the author at *mnh@bu.edu*.

**Figure 8.1.**   Fish hook and contestants at the starting line of the Peak Performance Design Competition. (*Photo courtesy of Boston University Photo Services.*)

You watch a match between Fish Hook and another wedge-shaped vehicle that has a shape similar to yours but is less massive. You mentally nickname this second vehicle "Lite Wedge." The designers of Lite Wedge have sacrificed wheel torque and raw mass in favor of faster speed. The judge signals the start, and several seconds later, both vehicles arrive simultaneously at the top of the hill. They engage in a bumper-to-bumper duel, but Lite Wedge does not have enough power to dislodge the front end of Fish Hook.

Although Fish Hook holds its ground, it cannot make any additional headway because it's been designed to turn off power once its hooks have been set into the carpet. As shown in Figure 8.2, both vehicles are virtually equidistant from the centerline at the end of the fifteen-second time interval. The judge measures the car positions carefully with a ruler and declares Fish Hook to be the winner by a scant few millimeters. You're on the schedule later to run against Fish Hook, and you begin to question the efficacy of your wedge-shaped design. Will your slow moving vehicle be able to dislodge Fish Hook's front end and allow you to overtake the top of the hill?

Before your first match can be held, two other cars must compete on the ramp. One of them, shown on the left in Figure 8.3, is based on a design that is wedgeshaped but looks nothing like yours. Nicknamed "Molly" by its creator, this vehicle has a wedge made from a small and narrow aluminum plate. Molly has large wheels that are completely exposed to her opponent's offenses. Molly's designers have added short spikes to her wheel treads in the hope of achieving better traction. Molly's opponent, shown on the right in Figure 8.3, resembles an army tank. It uses caterpillar treads and is built on a completely exposed chassis. Molly easily dislodges "The Tank" even though the latter is first to arrive at the top of the hill. Tank's designers have included no defense mechanism to prevent the vehicle from being pushed backwards down the ramp, and it cannot maintain its position.

**Figure 8.2.** The judge carefully measures the positions of Fish Hook and Lite Wedge at the top of the hill. (*Photo courtesy of Boston University Photo Services.*)

Your first match places you in a run against a vehicle nicknamed "The Scamper." The Scamper is an extremely fast, lightweight vehicle that zooms to the top of the hill in a few seconds while leaving behind a weight and string. When the trailing string is pulled taut against the jettisoned weight, it helps to slow the vehicle and stop it at the top of the hill. The taut string also releases a spring-loaded nail at the rear of the vehicle. This nail is designed to prevent the vehicle from being pushed backwards down the ramp.

As shown in Figure 8.4, the lightweight Scamper is no match for your massive, wedge-shaped vehicle. You easily dislodge your opponent at the top of the hill, even though you arrive there several seconds later. Your vehicle is strong enough and your wedge angle steep enough that you lift up Scamper and raise the tip of its nail right off the carpet. Your stopping mechanism places you precisely at the top of the hill and you're declared the winner of the run. Your strategy of combining a low gear ratio, slower travel speed, and higher wheel torque with the mechanical advantage of a wedge shape has paid off.

The next match places Molly against Fish Hook. The tip of Molly's narrow wedge catches under several of Fish Hook's barbs and dislodges them from the carpet. Molly's wedge is too narrow to reach them all, however, and some of the hooks retain their grasp, allowing Fish Hook to hold its position at the top of the hill. As you watch this race in action, an idea comes to mind. You decide to try adding a thin metal strip across

**Figure 8.3.** Molly's narrow wedge approaches Tank who has arrived first at the top of the hill. (*Photo courtesy of Boston University Photo Services.*)

**Figure 8.4.** Your strong, wedge-shaped vehicle easily dislodges the lightweight Scamper. (*Photo courtesy of Boston University Photo Services.*)

the entire leading edge of your wedge-shaped chassis. This strip, which protrudes about 2 cm from your wedge's front end like a blunt knife, just grazes the carpet as your vehicle travels along. Hopefully, it will be able to dislodge all of Fish Hook's embedded hooks. You've designed your modular vehicle to accommodate rapid changes, as permitted between runs by competition rules, so the change is easy. You add the metal strip just before your run against Fish Hook. Your wide metal strip dislodges all of Fish Hook's hooks, and you dominate the top of the hill. You've made it past a formidable foe.

You watch as the wedge-shaped vehicle designed by Tina and Juan, dubbed "Tijuana," competes against several other vehicles. Despite its clever harpoon design, Tijuana's stopping mechanism does not work as its designers had planned. In using a single switch placed beneath the chassis, Tina and Juan have failed to account for the height of the carpet pile. The carpet that covered their own test ramp had a much shorter pile than does the carpet covering the Peak Performance Design Competition ramp. Their stopping switch keeps snagging the ramp where its slope changes from upward to horizontal, causing their vehicle to stop short of the top of the hill. Their harpoon design works well, however, and it prevents several opponents from coming anywhere near the top of the hill. Tijuana wins several matches. Between runs, Tina and Juan solve their switch snagging problem by moving their rear wheels forward, thereby decreasing the axle-to-axle wheel span of their vehicle and leaving more room for the switch as it rounds over the transition point in the ramp. As you have done, they've designed a vehicle that is easily modified.

Tijuana next competes against Scamper. Scamper is so fast that it races to the top of the hill before Tijuana's harpoon can be launched. Tijuana is a strong wedge, however, and it easily dislodges the lightweight Scamper from the top of the hill and wins the run.

You finally are paired against Tina and Juan and prepare for a wedge-against-wedge contest. As the judge bellows out, "One, two, three, go!" you engage your starting mechanism and Tina presses her starting switch. Soon after the start, Tijuana's harpoon fires, firmly lodging itself in the carpet in front of your path. You are dismayed to see the harpoon work so well, because you assume that your vehicle will not be able to dislodge its well-placed barbs. When your vehicle reaches the embedded harpoon, your newly installed knife edge dislodges the harpoon! The unseated harpoon rises over your vehicle and you continue toward the top of the ramp. You meet Tina and Juan's vehicle face to face at the top of the hill, and you each prepare for a battle of the titans. You watch as your well-balanced, smooth-framed vehicle burrows under theirs, lifting it off the ramp! When Tina and Juan moved their rear wheels forward, so that their stopping switch could clear the higher carpet pile, they shifted their center of gravity toward the rear of the axle supports. This change has provided your wedge with more mechanical advantage than it might otherwise have had. You lift their front wheels off the ramp, push their wedge backwards a few centimeters, and capture the top of the hill. Your flexible design and well-conceived midrace design modifications have led to victory.

# Index

**SYMBOLS**

(HTML) code 107
"dip" chips 101
"ground" node 137
"idea chaos." 64

**NUMERICS**

10-speed bicycle 161

**A**

A switch pole 147
A/D converter 153
Abiomed 81
accelerometers 115
Accreditation Board for Engineering
    and Technology 196
actuator 115
Aegis air-defense system 191
aeronautical 2
aeronautical engineer 2
aerospace engineer 2
agricultural 2
agricultural engineer 3
AIAA 11
air bearings 105
airbags 115
American Institute of Aeronautics and
    Astronautics 11
American Institute of Chemical
    Engineers 12
American Institute of Chemical
    Engineers (AIChE) 12
American Rocket Society 11
American Society of Civil Engineers
    (ASCE) 12
American Society of Mechanical
    Engineers (ASME) 14, 205
amplitude modulation 149
analog-to-digital (A/D) 139, 141, 153
analog-to-digital/digital-to-analog
    conversion 153
analysis 23
    distinction between 23
Analysis Tool 148
angle irons 67

Ansoft™ 107
anthropometric 157
Apollo 13 179
armature 147
artificial heart 81
ASME 14
assembly drawing 131
Assembly language 140
Association for Computing Machinery
    (ACM) 13
asynchronous link 152
Athlon™ 140
ATM SIMULATOR 215
AutoCAD™ 36
Automated Teller Machine
    Simulator 215

**B**

Bad Engineer 37
bad engineer, vs. good engineer 37
Basic Stamp™ 140
batteries 39
battery 91
battery charger 165
BAUD 152
beta test 37
Big Dig 4
bin histogram 175
binary number 141
binary-weighted encoding 153
bioengineer 3
bioinformatics 4
biomedical 2
biomedical engineer 3
Biomedical Engineering Society 11
Biomedical Engineering Society
    (BMES) 11
bits audio 152
block diagram 149
body 204, 206
Boolean algebra 140, 153
brainstorming 63–73, 87
    formal 64
    ground rules for 64
breadboard 100
budget 150

bug 17, 182
bugs 36
build 35, 38, 44
Build Document and Test 35
buoy 23
burn-in 37

**C**

C 107
C++ 107, 117, 126
CAD 79, 128, 132
capacitors 137, 148
case studies
    De Haviland Comet 192
    Hartford Civic Center 183
    Hubble Telescope 192
    Kansas City Hyatt 187
    Space Shuttle *Challenger* 185
    Tacoma Narrows Bridge 183
    Three Mile Island 189
    *USS Vincennes* 191
CD 105
CD stamper disks 105
Cell phone 179
cellular telephones 6
center of mass 121, 123
Central Artery/Tunnel Project 4
CEO *see* chief executive officer
CEREAL DISPENSER L 171
chemical 2
chemical engineer 4
chemist 16
Chernobyl 189
chief executive officer 18
choice map 41
choose 38
chuck 105
civil 2
civil engineer 4
cognition 158, 159
commas 222
Common Writing Errors 222
communication
    effective 195–226
    oral 196
    written 196

communication skills 196
commutator 147
compass 170
compilation phase 66
composites 199
computer 2, 148
computer aided design 128
computer-aided design (CAD)
 7, 128, 181, 182
computer engineer 6
computer keyboard 157
computer numeric control (CNC) 132
computer scientist 6
Computer-Aided Drafting 128
computer-aided manufacturing 7
computers
    as analysis tool 148
    part of the design solution 154
conferences, preparing for 196
conversion algorithm 153
corona discharge 106
cost 20
cost requirements 31
cross section 134
Curie, Marie 16

**D**

data 23
Data Protocol 224
dc motor 48, 147
D-cell battery 104
De Haviland Comet 192, 193
Defense Arpanet 6
define 38
defining design objectives 33
DeMorgan's theorem 154
design 22, 23
    definition 22
    distinction between 23
    example 38–49
    good versus bad 29, 29–32
    objectives 40
    use of the "word" 23
design cycle 33, 34, 36, 38, 40, 50
    build 35
    document 35
    first cut 35
    information, gathering 33
    objectives 33
    revise 36
    test 35
    test finished product 37
Design Example 38
Design Strategy 33

design strategy 35, 41
    choosing 33
design, analysis 23
differential 44
differential equations 135
DIGITAL CLOCK K 165
digital signal 142
digital voice recorder 169
digital-to-analog 139
digital-to-analog (D/A) 153
Digital-to-Analog Conversion 141, 153
dimensioned sketch 146
Dimensioning 146
diode 138
diodes, transistors 137
director of marketing 18
document 35, 38, 44
documentation 36, 51, 74, 88
documentation trail 74
dot product 92
double-pole switch 147
drafting 128
drawings 79
DTMF (dual-tone, multiple-frequency,
    or Touch-Tone™) 159
dual-inline packaged 101
duty cycle 152
dynamic behavior of a system 135
dynamic system 136

**E**

Edison, Thomas 18
EEPROM 141
Einstein, Albert 16
elastic constant 112, 200, 202
elastic limit 111, 200, 201, 202, 203
electrical 2
electrical circuit 136
electrical energy 50
Electrical Engineer 6
electrical engineering 6
Electrical Power 92
electrical power 49, 50
electrically-erasable 141
electromagnetic actuator 189
electronic circuit 137
electronic documentation 75
electronic mail 195, 204
    writing 203–208
Electronics Workbench™ 36
electrostatic 105
electrostatic powder coating 94
e-mail 204
embedded computing 140

Energy 92
engineer's logbook 76, 77
engineering 21
    defined 1–21
    skills 19–21
engineering design failure 182
engineering design tools 60–89
    brainstorming 63
    documentation 74
    project management 83
    teamwork, as a design tool 61
engineering drawings 128
Engineering Notebook 79
engineering, fields of 2
    aeronautical engineer 2
    agricultural engineer 3
    biomedical engineer 3
    chemical engineer 4
    civil engineer 4
    computer engineer 6
    electrical engineering 6
    industrial engineer 6
    mechanical engineer 8
    mechotronics engineer 9
    naval engineer 9
    petroleum engineer 9
engineering, professional organizations 10
    American Institute of Chemical
        Engineers (AIChE) 12
    American Society of Civil Engineers
        (ASCE) 12
    American Society of Mechanical
        Engineers (ASME) 14
    Association for Computing Machin-
        ery (ACM) 13
    Biomedical Engineering Society
        (BMES) 11
    Institute of Electrical and Electron-
        ics Engineers (IEEE) 13
    Institute of Industrial Engineers
        (IIE) 14
engineering, skills
    experience 20
    intuition 20
    knowledge 19
engineers' logbook 75, 76
engineers' notebook 75
English bike 161
Ergonomic 174
ergonomic design 160
Ergonomic Measurements 174
ergonomics 20, 156, 156, 157, 158
estimation 91, 144
ethics 32
Excel™ 121

experience 19, 20, 21
   seasonsing 20
exploded view 131
exponential equation 137
eye contact 198

**F**

failure 20, 181
   in engineering design 194
   preparing for 193
Failure case studies 182
Failure in Your Own Design 193
Fatigue Tests 199
fictitious company 16
finite element analysis 132
first cut 35, 38, 44
First Cut at Design 35
First Sentence 205
first sentence 204
Flash Gordon 179
flowchart 143, 148, 149
force (weight) 92
Force-Displacement 111
Formal Memos 208
formal presentation 198
Fortran 107
front-wheel drive 44
furnace 135

**G**

Gantt chart 83, 84, 85, 89
garbage in 120
garbage out 120
gather 38
Gather Information 33
gear box 48
genomics 4
Glitches 36
Good Enginee 37
good engineer vs. bad engineer 37
Good Technical Documents 219
GPIB (general purpose instrument
   bus) 139
graphical programming 139
graphical user interface 103, 159
graphics 128
gravitational 92
green manufacturing 7

**H**

hand gestures 197
hand sketches 130

Hand-set TELEPHONE DIAL L 159
harpoon 68, 69, 108, 112, 231
Hartford Civic Center 183
Hawking, Steven 196
header 204
heat sinks 146, 147
heat-transfer 147
Hewlett-Packard instrument bus 139
hip replacement joint 102
histograms 175
Hoover Dam 4, 5
Hopper, Grace 17
HP-VEE™ 139
HSPICE 137
HTML (hypertext markup language) 120
Hubble space telescope, 61
Hubble telescope 192
human-machine interfaces 155, 191
Hyatt hotel 187
Hyatt Regency walkway collapse 187
hydride 165

**I**

idea purge phase 65, 67
idea trigger method 64, 65
idea trigger phase 65
Identification
   Friend or Foe 191
Identification Friend or Foe (IFF) 191
idler wheels 72
IEEE 13
IEEE-488 bus 139
IIEs 14
image processor 129
imaging 129
inductors 137
industrial 2
industrial engineer 6
infinitives 223
Informal Brainstorming 69
informal meetings 197
   preparing for 196
information, gathering 33
Institute of Aerospace Sciences 11
Institute of Electrical and Electronics
   Engineers (IEEE) 13
Institute of Industrial Engineers
   (IIE) 14
Instruction Manual 213
instruction manual, preparing 213
integrated circuits 101
International Space Station, 61
Internet 6, 120
intuition 19, 20

ion machining 105
isometric projections 130
iteration 41

**J**

Java 107
Johan Vaaler 173
journal 212–213
   articles, writing 212
   paper, writing 213
Journal Articles 212
Journal Paper 213

**K**

Kansas City Hyatt 187, 189
kernel 103
kinetic energy 110
Kirchhoff's current law 146
Kirchhoff's voltage law 146
knowledge 19
knowledge tools 90

**L**

lab technician 18
LabVIEW™ 139
laser 8
Laser Printer 180
laws of time estimation 86
lawyer 18
letters, writing 208–211
light-emitting diode 165
light-emitting diode (LED) 137
LIGHTING SYSTEM L 167
linear 137
linear material 200
load cell 200
Loading Tests 199
logbook 122
   engineer's *see* engineer's logbook
   example 77
   format 76
Lotus 121

**M**

Magnetic resonance imaging 179
maintenance 31
manufacturability 20
manufacturing engineer 6
MARINER'S COMPASS J 170
marketability 20
MathCad™ 107

Mathematica™ 107
mathematician 17
MATLAB 112, 117, 126, 135
mechanical 2
mechanical advantage 48
mechanical engineer 8
mechanical load 139
MECHANICAL LOADING 199
mechanical loading tests 209
Mechanical Power 92
Mechano™ 67, 68
mechotronics engineer 9
Melville, Herman 68
memoranda 79
    writing 203, 208–211
memory 141
MEMS 10, 115
mentor 32
mercury 42
mercury switches 39
micro 6
micro-electromechanical 6
micro-electromechanical systems
    6, 115
micro-electromechanical systems
    (MEMS) 4, 6, 9
microprocessor 6, 140, 143, 144,
    149, 154
milling machine, drill press 97
Ming the Merciless 179
*Moby Dick* 68
modular approach 34
mountain bike 161
mousetrap 39, 148
MRI 179

**N**

Napster 179
NASA Space Shuttle 2
National Oceanic and Atmospheric
    Administration (NOAA) 23
Natural speech 197
naval 2
naval architect 9
naval engineer 9
Newton's law 109
Newton's law of motion 23
nickel-metal 165
NONLINEAR CIRCUIT
    SIMULATION 138
normally closed 70, 71
normally open 70, 71
nuclear magnetic resonance 179

**O**

objectives 33
OEM 164
Ohm's law 137, 146
On-the-job training 20
open circuit 180
operational amplifiers 137
optical detector 142
oral communication 196
organizational chart 83
original equipment manufactured 164
original equipment manufacturer 33
O-rings 185
orthographic projection 130

**P**

Palm Pilot™ 180
PAPER CLIP J 173
parallel link 151
parallelism 222
past, present, and future tense 222
PDF (portable document format) 120
Peak Performance 108, 121
peak performance design competition
    51, 223, 227
Pentium® 140
permittivity 115
perspective drawings 130
petroleum 2
petroleum engineer 9
Phase 65
photonics 6
physicist 16
ports 141
potential energy 92, 110
powder-coating 94
Powell, Colin 18
Power 91
power 146
    transistors 146
power source 42
power-flow 91
prime mover 42
Princess™ telephone 160
principal cross section 132
Prism Corporation 105
Product design review 197
product safety 31
production manager 18
ProEngineer 132, 135
project management 15–19, 83–85, 89
Project status review 197
Prony brake 49
Proposals 212

proposals 212, 213
    writing 212
propulsion mechanism 41, 42
prototype 35, 104
PSPICE™ 36
pulse width modulation 152
Punctuation Around Parentheses 223
purge-trigger sequence 65

**R**

ramming device 68
rapid prototyping 132
reaction time 176
Real-Time Computer Control 151
real-time control 151
recipient 204
reduction to practice 75
reliability 20
reproduction 23, 23, 36
    distinction between 23
resistors 137
restoring force 116
resultant vector 144
reverse engineering 148
revise 36, 38, 44, 45
revision cycle 37
REVOLVING DOOR J 172
robotics 7
Roosevelt, Eleanor 18
rotor 147
rubber bands 39

**S**

sacrificial layers 115
safety margins 188
scale modeling 102
scale models 4
schematics 79
seasoning 20
sender 204
serial link 152
short circuit 180
significant figures 96, 97
Significant Figures, Dimensioning, and
    Tolerance 96
Simulation 107
Simulink® 135
Simulink™ 36, 107, 139
single-pole, double-throw (SPDT)
    switch 147
snap through 117
software documentation 79
solid model 132

solid models 130
SolidWorks™ 128
space frame structure 183
Space Shuttle 185
Space Shuttle *Challenger* 185, 192
SPEED CONROL 142
SPICE 137, 138
spreadsheet 122, 124, 125, 150
spring constant 110
SQUEEZE KETCHUP
    CONTAINER J 171
Star Trek 179
start bit 152
starting device 41, 43
static electricity 106
stator 147
stop-bit 152
stopping mechanism 41, 42
stopping switch 231
strain 199, 200
strain gage 200
strategy for producing good technical
    documents 219
stress 199, 200
Stress-Strain 199
student, gaining experience as a 21
subject 204
supercomputers 9
surface micromachining 115
switch 70, 147
switch pole 147
synchronous link 152
systems 2
systems engineer 10

**T**

Tacoma Narrows Bridge 183
tape cassette 164

team building 61–63
teamwork, as a design tool 61–63
technical drawing 128
Technical Reports 212
technical reports 79, 212
    writing 212
tensile test machine 200
test 35, 38, 44
test finished product 37
The rotary telephone dial 160
thermal flow 147
thermal memory 135
thermal resistance 147
thermostat 135
Thomas, Clarence 18
threaded rod 67, 72
Three Mile Island 189, 190
thyristors 146
time constant 135
time line 84
tolerance 97
tolerance table 97
toner particles 180
torque 49, 147
Touch-Tone® 160
traction 44, 228
trajectory 108
transcutaneous energy transfer
    device 81
transformers 137
transistor 137
troubleshooting 214
TSPICE 137

**U**

U.S. National Oceanic and Atmospheric
    Administration 23
Use of the Word "design" 23

Use of the Word "This" 222
Use of the Words "Input"
    and "Output" 223
USS Vincennes 191

**V**

vanishing point 130
Vecor manipulation 144
vector 144
Vectors 144
Vincennes 191

**W**

Washington, George 18
weather buoy 24
Web 195
wedge-shaped design 227
wedge-shaped vehicle 228, 231
witnessed and understood 77
World Heart 81
World Wide Web 6, 121
writing
    journal articles 212
    letters 208–211
    proposals 212
    technical reports 212
writing errors
    commas 222
    infinitives 223
    parallelism 222
    past, present, and future tense 222
written communication 196

**Z**

zinc air batteries 68